Edited by James Warren

Managing Transport Energy

Power for a Sustainable Future

Oxford University Press in association with The Open University

Published by Oxford University Press, Great Clarendon Street, Oxford OX2 6DP in association with
The Open University, Walton Hall, Milton Keynes MK7 6AA

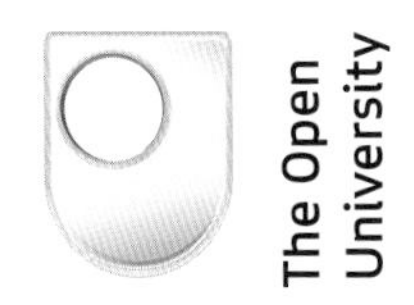

Oxford University Press is a department of the University of Oxford. It furthers the University's objective of excellence in research, scholarship, and education by publishing worldwide in

Oxford New York Auckland Bangkok Buenos Aires Cape Town Chennai Dar es Salaam Delhi Hong Kong Istanbul Karachi Kolkata Kuala Lumpur Madrid Melbourne Mexico City Nairobi São Paulo Shanghai Taipei Tokyo Toronto

Oxford is a registered trade mark of Oxford University Press in the UK and in certain other countries.

Published in the United States by Oxford University Press Inc., New York

First published 2007

Edited and designed by The Open University

Typeset by S R Nova Pvt Ltd, Bangalore, India

Printed in the United Kingdom by The University Press, Cambridge.

This book forms part of an Open University course T206 *Energy for a sustainable future*. It also includes two other copublished books (*Energy Systems and Sustainability*, ISBN 0-19-926179-2 and *Renewable Energy*, ISBN 0-19-926178-4), study guides, and other supplementary print and media material.
Details of this and other Open University courses can be obtained from the Student Registration and Enquiry Service, The Open University, PO Box 197, Milton Keynes, MK7 6BJ, United Kingdom:
Tel +44 (0)870 333 4340, email general-enquiries@open.ac.uk
http://www open ac uk

British Library Cataloguing in Publication Data available on request

Library of Congress Cataloging in Publication Data available on request

ISBN 978 0 1992 1577 5

1 3 5 7 9 10 8 6 4 2

About the authors

Marcus Enoch

Dr Marcus Enoch is a lecturer in transport studies at Loughborough University (UK), where he has worked since January 2003. Prior to this he was a research fellow at the Open University (1999–2002) and before that, a transport journalist for the practitioner publication *Local Transport Today*. He has a wide range of research interests in the field of transport planning policy and has written widely about such diverse issues as parking, buses, demand-responsive transport systems, car clubs, travel planning, funding mechanisms for public transport and the use of the tax system for influencing modal shift. His work has also investigated the development and application of transport policy in island nations.

Ben Lane

Dr Ben Lane is director of Ecolane Transport Consultancy, a company he established in 1996. His career has spanned both academic research and independent consultancy in both public and private sectors. Ecolane's projects have ranged from life cycle assessments of cleaner vehicles to the provision of travel plan training. He is particularly interested in the role of consumer attitudes and believes that low carbon options can only be effectively promoted once an understanding of the psychological factors that influence vehicle purchasing is achieved.

Stephen Potter

Stephen Potter is Professor of Transport Strategy at The Open University (UK), undertaking work on the design processes involved for the diffusion of cleaner transport technologies and the development of sustainable transport policies. This also includes work on travel planning and the design of environmental taxation on transport. He also undertakes studies in other areas related to sustainable design. Professor Potter works in both the Design Innovation Group and in the Energy and Environment Research Unit (EERU) and is the Director of the Centre for Technology Strategy; much of the University's postgraduate training is delivered through this unit.

James Warren

Dr James Warren is an academic in the Faculty of Technology at The Open University; his work focuses on transport systems and in particular emissions control, emissions modelling and motorised mobility. This includes work in the area of greener designs for supply chains. He has collaborated with all the co-authors.

Contents

Introduction

by James Warren

Introduction

The transport sector accounts for 36% of all energy used in the UK. Transport energy in this context includes all land, air and sea vehicles but is mainly dominated by fuel use for land transport. The significant level of energy use for this purpose is repeated in many locations in the world and the UK is typical of many European countries. For the majority of developed countries in the world, the energy used for transport is predicted, up to 2030, as growing by nearly 2% every year (IEA, 2002 and DTI, 2005a, 2005b).

This level of intensive energy use for mobility is also replicated in the United States of America; for example, in terms of transportation energy use, the USA's demand for travel consumed an impressive 28% of all energy in 2005. This has increased by about 4% every decade from 1985 when it was 20% (Davis and Diegel, 2005). In 1974, in the USA consumption was about 18% of total energy demand, and for the long-term future is set to grow. By the year 2030, the USA will be consuming more than 25 million barrels of oil per day, with transport taking about 20 of those. Current consumption is just under 14 million barrels per day (Davis and Diegel, 2005; EIA, 2006).

In terms of petroleum-derived products, such as petrol (gasoline) and diesel, transport's share was approximately 65% in the UK during 2004 (DTI, 2005b) and this share is growing slowly and steadily with time. Figure i.1 shows global oil consumption in all sectors (transport, industry, other and electricity generation combined) and global transport energy demand for all fuels (including oil). The growth in demand for energy by the transport sector can be clearly seen. The share of liquid fuels used by transport is predicted to grow by approximately 7% from the present day (2006) to 2030 (IEA, 2002).

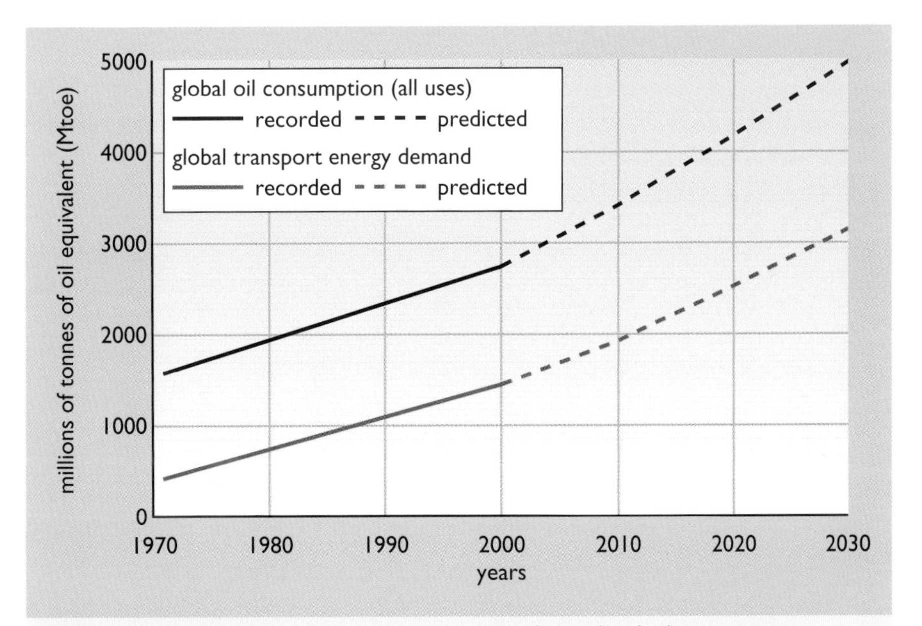

Figure i.1 Growth in worldwide transport energy demand and oil consumption (adapted from IEA, 2002)

Understanding the various issues that affect how and where energy is utilised for transport, and the possible options that are on offer to reduce energy use in the transport sector, is the main aim of this book. The setting for much of the book can be observed in the world around us – every day we are bombarded with reminders of some of the following issues:

- the premise that fossil fuel prices may rise to unaffordable limits, disrupting economies
- increasing concerns about negative environmental impacts
- the growing awareness of negative social impacts such as climate change, poorer local air quality, congestion, accidents and injuries, and generally less healthy lifestyles as a result of our reliance on transport.

However, the biggest concern is that fossil fuels are finite and will run out in the short to medium term. The prospect of yet more oil crises and further increases in carbon dioxide levels eventually (if not already) destabilising the earth's environmental systems, is now beginning to seem all too real. Predictions have varied as to when oil will run out; these calculations are highly dependent on how reserves are estimated and on estimates of future consumption. Nevertheless, world production of liquid fossil fuels from all sources is predicted to peak before 2015, and the production peak for natural gas is expected to occur around 2030 (Campbell, 1997; Campbell and Laherrère, 1995, 1998; BP, 2003).

All of these driving forces have combined to raise awareness about the importance of reducing energy consumption in all sectors, particularly in the area of transport. In the UK, nearly two-thirds of all petroleum supply is used in transport (DTI, 2005b). It is clear that fossil fuels dominate in the world energy market and that for transport in particular, oil-based products remain the largest source of energy.

For the majority of people, our reliance on personal car use remains the biggest issue in realising a more sustainable transport lifestyle. This book explores many of the issues and potential solutions for ultimately reducing transport energy use.

The structure of this book

In this book, the effects of transport are reviewed by focusing on the key factors which offer ways of decoupling transport from intensive energy use. Topics covered include a thorough treatment of fuel efficiencies, occupancy, population effects, and overall mobility levels (Figure i.2). By examining each of these factors in turn, it is possible to analyse various scenarios and so model what is likely to happen for any given set of parameters.

Chapter 1 carefully addresses the issue of what is really meant by sustainable transport from the individual's perspective and explains why personal transport consumption is such a major part of the overall energy demand.

The effects of technology have a significant role to play in the reduction of transport energy and understanding these technologies is critical to modelling how they may contribute to this in the future. The issues examined in Chapter 2 include internal combustion engines and transport

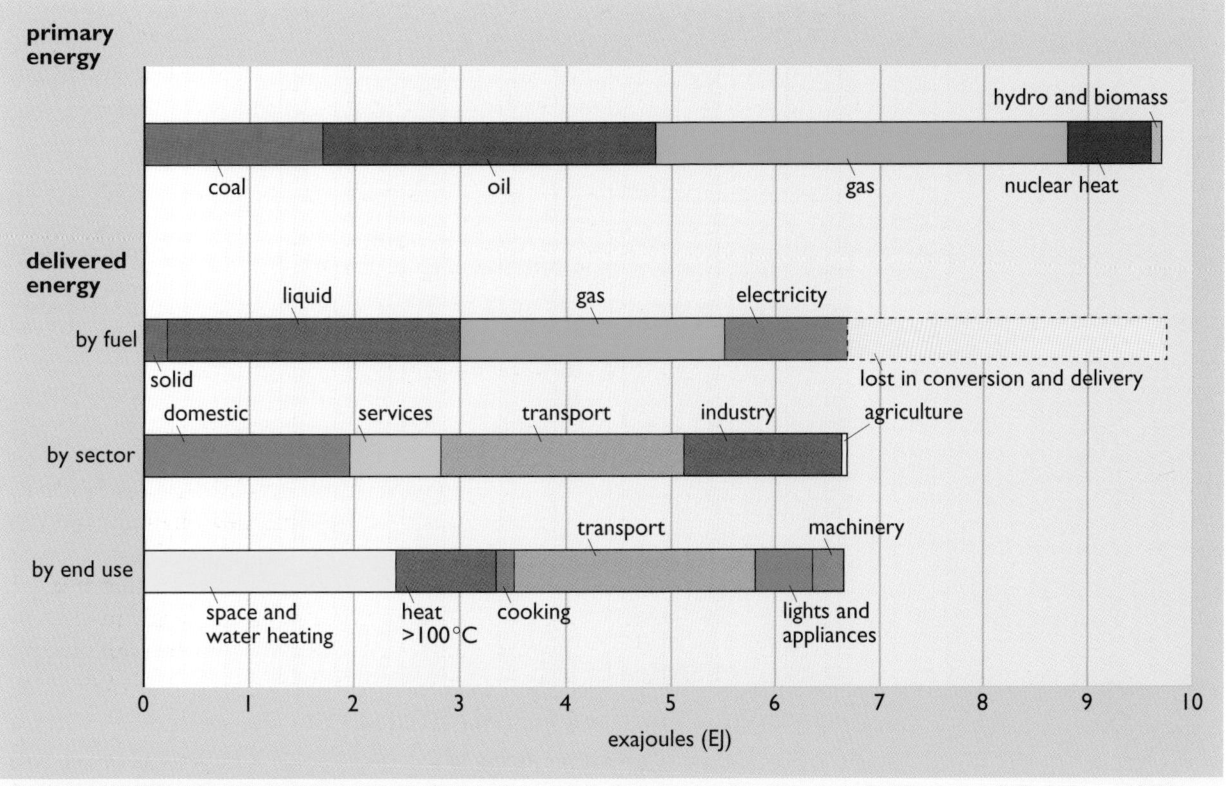

Figure i.2 UK primary energy by fuel, and delivered energy by fuel, sector and end use, in 2000 (source: DTI, 2001a, 2001b)

fuels of all types, with some regard to the introduction of future fuel systems. Changes and innovations in transport power supply, such as hybridisation and electrification, are explained along with a discussion of the types of renewable fuels that are being considered in order to meet the demand. These can all be thought of as a technological fix and although they do indeed result in significant energy savings there is always a chance that rebound effects will incur further consumption.

A major portion of Chapter 3 is devoted to mobility management. This is a key concept in managing the demand for transport and the text explains the dilemma between increasing mobility and the need for curbing externalities resulting from our use of transport. Travel demand, travel blending and travel plans are defined and illustrated with pertinent case studies.

Travel plans are discussed in more depth in Chapter 4, with an emphasis on personal transport as well as freight. Developing travel plans is depicted in detail using various examples, including case studies from business, education, health care, industry and residential sites. The measures that are required to plan travel changes and then bring them about are illustrated. Sustaining travel plans is also contemplated.

The Conclusion offers advice on the way forward to 'greening transport' using the model set out in the earlier chapters, which illustrates the application of a wide variety of measures intended to reduce overall energy consumption in the transport sector. An overview of behavioural changes, technologies, and managing and planning travel, along with some consideration of secondary impacts from transport, is considered in order to lower overall energy consumption in this sector.

References

BP (2003) *BP statistical review of world energy* [online], BP, http://www.bp.com/centres/energy2002/index.asp [Accessed 22 May 2006].

Campbell, C. J. (1997) *The Coming Oil Crisis* , Multiscience Publishing and Petroconsultants S.A.

Campbell, C. J. and Laherrère, J. H. (1995) *The World's Supply of Oil, 1930–2050*, Geneva, Petroconsultants S.A.

Campbell, C. J. and Laherrère, J. H. (1998) 'The end of cheap oil', *Scientific American*, March, pp. 59–65.

Davis S.C. and Diegel S.W. (2005) *Transportation Energy Data Book* (25th edn), Tennessee, Oak Ridge National Laboratory.

Department of Trade and Industry (DTI) (2001a) *Digest of UK Energy Statistics (DUKES), 2000*, London, HMSO.

Department of Trade and Industry (DTI) (2001b) *UK Energy Sectors Indicators*, 2001, London, HMSO.

Department of Trade and Industry (DTI) (2005a) *Digest of Energy Statistics*, London, The Stationery Office. ISBN 011 5155 139.

Department of Trade and Industry (DTI) (2005b) *Energy in Brief*, London, The Stationery Office; also available from http://www.dti.gov.uk/energy/statistics/stats-publications/UK%20Energy%20in%20Brief/page17222.html [Accessed 22 May 2006].

International Energy Agency (IEA) (2002) *World Energy Outlook 2002*, (2nd edn), Paris, France, OECD/IEA; also available from http://www.iea.org/textbase/nppdf/free/2000/weo2002.pdf [Accessed 22 May 2006].

Energy Information Administration (EIA) (2006) 'Annual Energy Outlook 2006 with Projections to 2030', report no.: DOE/EIA-0383(2006) [online], Energy Information Administration, Department of Energy, United States Government, http://www.eia.doe.gov/oiaf/aeo/index.html [Accessed 10 July 2006].

Chapter 1

Sustainability, energy conservation and personal transport

by Stephen Potter

1.1 Technical and consumption factors in transport's environmental impacts

Over the last 40 years, transport in the UK and in other developed economies has moved from being a relatively small consumer of energy to become the largest and fastest expanding energy sector. About 80% of transport energy in the UK is consumed by motor vehicles, and three-quarters of that is by cars, so *personal* transport is a major part of overall transport energy use and emissions. Our lives have become increasingly transport-dependent with, among other things, road congestion growing to unprecedented levels (Figure 1.1). The issue of what should be done to address the transport crisis has become a high-profile and highly contentious subject. This is not surprising, as car use is now accepted as normal and restrictions upon the 'freedom to drive' are much resented. Yet transport produces major local and global pollutant emissions, with a whole host of other transport-related issues such as accidents, and increasingly sedentary lifestyles leading to adverse health effects and obesity, together with social exclusion and the much-publicised problems and cost of congestion.

Figure 1.1 Traffic congestion on the Paris Périphérique urban motorway. Traffic congestion is now a growing part of everyday life, not just in city centres but almost everywhere in developed economies. Road building has failed to cut congestion, leading to policies to try to manage the amount of traffic

When global environmental concerns about transport first emerged in the 1980s, the initial reaction was to separate out environmental impacts from all the other issues of high car use. At that time, emissions from industry and production were viewed as the dominant concern, and for vehicles there was an initial emphasis on reducing emissions from vehicle production, with the automotive industry adopting the use of water-based paints, eliminating chlorofluorocarbons (CFCs) and adopting other clean production and pollution abatement technologies. However, by the early

1990s, environmental life cycle analysis (see Chapter 2) had established that the fuel consumed by motorised vehicles represents 80–90% of total life cycle energy consumption (for example, Teufel et al., 1993; Hughes, 1993 and Mildenberger and Khare, 2000).

This eventually led to a shift in focus from production to product design. Initially, local air pollution concerns resulted in the promotion of emission clean-up technologies, such as catalytic converters for car exhausts. In conjunction with this came moves towards cleaner fuels, including unleaded petrol, low sulphur diesel and the use of 'alternative fuels' such as liquefied petroleum gas (LPG), compressed natural gas (CNG), and electricity in battery-powered vehicles.

The agenda has now moved on from these air quality concerns, with a growing acknowledgement that actions are needed to address global environmental impacts, particularly CO_2 emissions from transport. The amount of CO_2 generated by transport in the UK has doubled in the last 25 years and transport is the fastest growing source of all emissions.

Road transport CO_2 emissions steadied in the 1990s, but have now started to rise again. The Department for Transport projects them to increase by 5% from 2005–10 and rise by double that rate between 2010 and 2015 (DfT, 2004a). Air travel, which is now a substantial source of CO_2 emissions, is

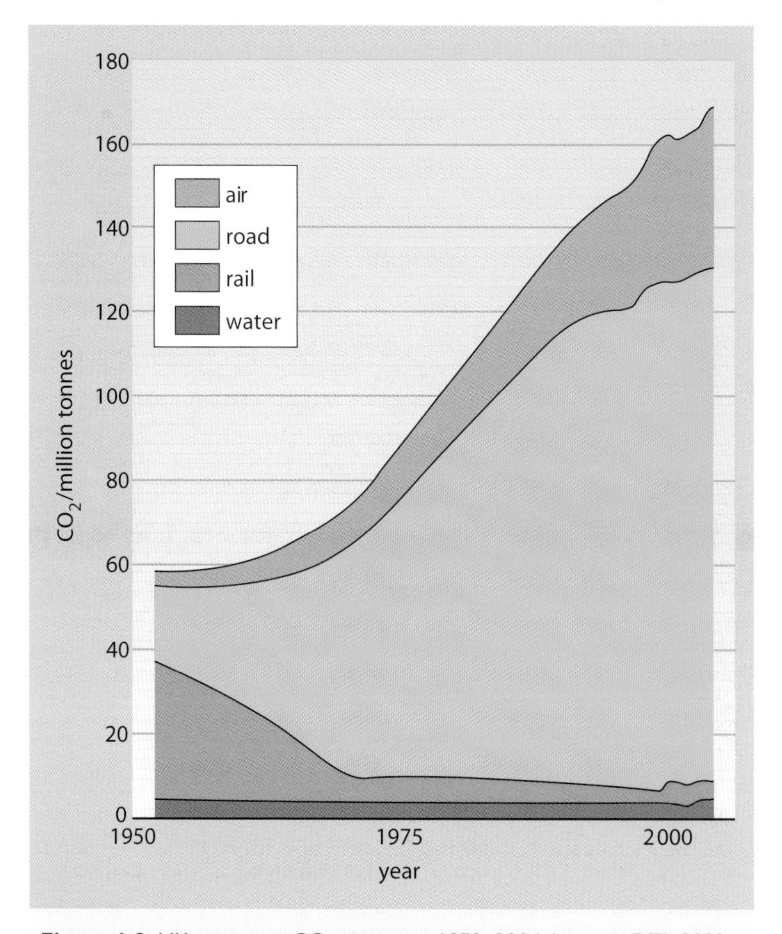

Figure 1.2 UK transport CO_2 emissions 1952–2004 (source: DTI, 2002 and Digest of UK Transport Statistics)

projected to grow faster than ground transport, albeit from a lower base. This book concentrates on ground transport, but there is a growing debate on aviation and sustainability (for example, see Bishop and Grayling, 2003).

The amount of CO_2 produced when a fuel is burned is basically a function of the mass of fuel consumed and its carbon content. A new emphasis has therefore arisen on fuel type, fuel economy in vehicle designs and the promotion of alternative fuels that have a lower carbon content.

In practice, vehicle fuel economy improvements may fail to make much of a difference, as increases in the amount of travel, and 'rebound effects', such as changes in drivers' and car buyers' behaviour, compensate for these. A classic example of such a rebound effect is shown by the *Corporate Average Fuel Economy* (CAFE) regulations in the USA. In the 20 years to the mid-1990s, these regulations improved car fuel economy by more than a third, but growing vehicle use has more than compensated for this, partly due to lower running costs arising from better fuel economy. Overall, although vehicle energy efficiency has improved, the total amount of fuel consumed (and therefore CO_2 emitted) has risen. In isolation, product energy efficiency measures do not always save energy, a point which has been made with reference to other energy sectors (see Herring, 1999).

The opposite extreme to vehicle efficiency improvements is the consumption-oriented view that behavioural change should be the main policy response to cut transport's environmental impacts. This implys a dramatic reduction in the use of the most energy-intensive transport modes of car and air travel. But, many individuals and politicians baulk at the prospect of 'turning the clock back' to a level of mobility considerably less than that we currently enjoy. Car use is now deeply entrenched in our society and economy, however environmentally problematic that may be.

Figure 1.3 The September 2000 UK fuel protests. The political sensitivity of transport was well illustrated by this direct action. When the price of oil led to high fuel prices there were calls for cuts in fuel taxation. Although organised by a very small number of people, the blockade of oil refineries rapidly caused transport chaos and the government quickly caved in, cutting over £1bn off fuel and lorry taxes. Environmental and transport policy issues were totally ignored. The cut in tax has subsequently led to a rise in fuel use and CO_2 emissions and the government has amended its CO_2 forecasts accordingly!

1.2 Exploring the issue

A combination of the two approaches (technical product efficiency improvements and changes in consumption or behaviour) would therefore appear appropriate, but what should be the relative contributions of each? (see Potter, 1998; Potter, Enoch and Fergusson, 2001; Potter and Warren, 2006). This chapter seeks to provide an overview of the sort of improvements that are needed in product efficiency, and the changes to consumption patterns that are needed to cut CO_2 emissions from personal transport to a sustainable level. The possible measures and technologies that could be used to achieve these improvements are then considered in detail in Chapters 2, 3 and 4.

A useful way to explore the role of technical and behavioural aspects is to investigate the key factors in the generation of transport's environmental impacts. This can be done in a 'backcasting' exercise to identify a future sustainable level of CO_2 emissions, and then explore how we might get to that state through a mix of technical developments and changes in travel behaviour. In contrast to 'forecasting', backcasting is not about predicting where current trends will take us, but is used in policy studies where an alternative future is envisaged in order to see if and how it is possible to achieve that future. One simple, but fruitful approach to backcasting has been suggested by Ehrlich and Ehrlich (1990), and developed by Ekins et al. (1992), in which environmental impact (*E*) is expressed mathematically as the product of population (*P*), level of consumption (*C*) and technology used (*T*). This formula is:

$$P \times C \times T = E$$

Using this approach, and looking at the world as a whole, an example might be to assume that in the next 50 years or so global population will increase by around 60% and consumption will at least double. So, if present environmental impacts are expressed as an index of 1.0, then the current or 'baseline' position using the Ehrlich/Ekins formula is:

$$P \quad \times C \quad \times T \quad = E$$

$$1.0 \times 1.0 \times 1.0 = 1.0$$

If population increases by 60%, then its index number would rise to 1.6 and if consumption doubles, its index number becomes 2.0. If the technology does not change (i.e. all energy production and energy use technologies produce the same amount of environmental impacts as today), the formula becomes:

$$P \quad \times C \quad \times T \quad = E$$

$$1.6 \times 2.0 \times 1.0 = 3.2$$

So, if there is no change in the environmental performance of technologies used, the overall environmental impact (*E*) increases more than threefold. This is simply a result of more people each consuming more goods and resources. In this scenario, to prevent an increase in environmental impact *T* has to be reduced to just over 0.3 ($1.0/(1.6 \times 2.0)$). This might be achieved by, for example, a threefold improvement in energy efficiency or the use of less-polluting fuels and technologies – or some combination of the two.

But this is just to stop environmental impacts getting worse! If, for example, a sustainability target suggests we need to halve current environmental impacts, then the figure for *E* has to be reduced to 0.5. This results in the need for an even bigger improvement in the 'Technology' part of the equation – down to 0.16. This represents the need for a very large improvement in energy efficiency and/or a major shift to non-fossil fuels.

This simple index model can forecast changes in environmental impacts, but can also be used to identify a backcasting target. In doing so, the index model identifies a crucial point; with world population rising and economic growth leading to higher levels of consumption, then very major improvements have to be made in our production and use of energy for there to be any hope of addressing the world's environmental crisis.

What would a transport version of this simple index model look like? Total travel could be broken down into key emission-generating factors, which will help explore the role of consumption and vehicle efficiency in cutting environmental impacts from transport. As noted previously, life cycle studies have established that the fuel consumed in driving vehicles represents some 90% of total life cycle energy consumption, so this is the issue upon which to concentrate. The Ehrlich/Ekins model can be developed to calculate future environmental impacts from motorised travel. For this, the baseline needs to be set at a specific year, say 2005. The formula can then be used to explore changes in *P*, *C*, *T* and *E*, by changing the values to represent hypothetical or reported trends that are expected in the future. These changes will all be represented as indexes relative to the baseline year. The indices of all the components at the baseline year are as shown in Table 1.1.

Table 1.1 Baseline Transport Emissions Index, set at 2005

Population	×	Car journeys per person	×	Journey length	×	Emissions per vehicle km	=	Total emissions
1.0	×	1.0	×	1.0	×	1.0	=	**1.0**

This provides us with a very simple, but nevertheless useful transport model. It allows us to look at how changes in the values of these four key components will affect the total emissions produced. Throughout this chapter we will use this simple index model to undertake a backcasting exercise, exploring possible transport futures with different technologies and policy approaches.

1.3 Business as usual

One way to use this index model is to concentrate on the key global issue of CO_2 emissions from personal transport. A reasonable timescale might involve looking ahead 20 years, since beyond this it is hard to envisage the changes that could occur in transport technologies and policies. So, what level might a transport CO_2 index for the UK reach by the year 2025 if we assume a continuation of current transport trends? With UK population change expected to be relatively small, this factor can be left out, but the other key factors are shown in Table 1.2.

Table 1.2 Key travel trends, based on a continuation of past trends, 2005–25

2005 data and current trends	Index by 2025
Car journeys average about 600 per person per year (currently rising by 14 per year)	1.5
Journey length averages 13.6 km (rising at about 0.14 km a year)	1.2
Fuel use averages 9.1 litres per 100 km across the UK car fleet (assumed to improve to 8 litres per 100 km)	0.9

Sources: Noble and Potter, 1998 and DfT, Transport Statistics Great Britain (editions to 2005)

The rate of fuel economy improvement described in Table 1.2 is faster than that achieved historically (which is only 0.2% a year). The 1996 European Union *Auto-Oil* voluntary agreement with the car industry has improved test fuel and CO_2 emissions, but not to the level hoped. Some EU states have a fuel economy considerably better than that of the UK, and the figure of 8 litres per 100 km has been taken as realistic for 2025 as it is the average level of fuel economy already achieved in the Netherlands and Italy. Equivalent fuel consumption for 9.1 litres per 100 km is 31 miles per gallon (or 26 miles per US gallon which is 20% smaller – see Box 2.3 in Chapter 2). An improvement to 8 litres per 100 km would mean a fleet average rise to 35 mpg.

For our 20-year timescale, a 'business as usual' (BAU) forecast could envisage the continued use of oil for personal transport (the use of alternative fuels will be explored later in another scenario). With the continued use of oil, the carbon content would remain the same, and so emissions would simply be a function of the amount of fuel used. Such a future would result in the formula becoming as shown in Table 1.3.

Table 1.3 UK business as usual (BAU) personal transport emissions index by 2025, baselined at 2005

Population	×	Car journeys per person	×	Journey length	×	Emissions per vehicle km	=	Total emissions
1.0	×	1.5	×	1.2	×	0.9	=	**1.6**

So, under BAU assumptions, CO_2 emissions will increase to 1.6 times their current level (i.e. a growth of 60%). This is looking only at the UK situation. Carbon dioxide emissions and global warming are, however, global issues and an isolationist approach that considers only CO_2 produced in the UK context is inappropriate.

Car ownership and traffic levels per capita in the developing world are growing much faster than in the UK. In 2001 there were about 700 million vehicles in the world of which 500 million were cars. These were heavily concentrated in the industrialised nations (in 2001, 15% of the world's population lived in OECD countries, accounting for more than 80% of car registrations).

Although car ownership is much higher in the developed OECD countries, car ownership and use in the developing world is growing at a very fast rate. The main increase in car use is now taking place in Eastern Europe and in the Asian economies of India and China. Between 1995 and 2002, the number of cars in Britain rose by about 2% a year; in Germany it was 1.5% and in the USA 1% (DfT, 2004a). All these countries have high existing levels of car ownership. In the UK there are about 50 cars per 100 people, in Germany 54 per 100 people and in the USA 80 per 100 people.

China has only 9 cars per 100 people but the growth in car sales is currently between 10 and 20% per annum – ten times the annual growth of the UK market. In 2004 China became the third largest car market, with sales exceeding 5 million cars. Within three or four years, China is set to overtake Japan to become the world's second largest car market and is forecast by 2010–15 to overtake the USA to become the world's largest car market (*The Economist*, 2005). Such car ownership growth is mirrored in other large developing countries such as in India and the countries of Latin America, where current ownership levels are relatively low.

Population is expected to rise by about 30% and the growth in car journeys is anticipated to be very large indeed. As noted above, the number of cars in the world is set to double, with the number of journeys expected to rise somewhat faster (giving a 2025 index figure of 2.3). These growth rates could be incorporated into a global BAU version of the index model (Table 1.4).

Table 1.4 Global business as usual (BAU) transport emissions index by 2025, baselined at 2005

Population	×	Car journeys per person	×	Journey length	×	Emissions per vehicle km	=	Total emissions
1.3	×	2.3	×	1.2	×	0.9	=	**3.2**

Overall, the result of all these trends suggests that CO_2 from personal transport could increase to more than three times the current levels within 20 years. The indices used are of necessity approximate. Global population growth may be less than 30% in 20 years, but other factors are likely to have been underestimated. The UK journey length and fuel economy indices are used because a global estimate is not available for either of these. In developing nations these other factors would have a strong role in pushing up CO_2 emissions, even were population growth to be lower than estimated.

The sheer rate of growth in car use in the developing world raises some difficulties distinct from environmental impacts. It seems likely that the world production of oil is near its peak. This growth has sustained the massive rise in car, air and freight transport in the developed world. Just as car use is taking off in Eastern Europe, China and the developing world, oil production is set to peak and start to decline. It is difficult to see how a growth in the numbers of petrol- and diesel-engined cars can be maintained for very much longer. Possibly, when oil production fails to meet growing demands, developing countries will be priced out as the developed countries secure their supplies. So, it is not just environmental impacts and emissions of pollutants that are unsustainable: the long-term availability of oil supplies is also in question. Current growth trends in car use seem to be both economically and environmentally unsustainable, with additional uncertain social, developmental and political implications as well.

This very simple exercise has major implications for any policy designed to address the CO_2 impacts of personal transport by improvements in vehicle energy efficiency alone. To hold total CO_2 emissions to their 2005 baseline level, the index figure for CO_2 emissions per vehicle kilometre would have to be drastically cut to compensate for growth in consumption and population.

For the UK, simply to stop the 2005 baseline total CO_2 emissions rising would require the emissions index to be cut to 0.56 (using Table 1.3: ($T = E/PC = 1.0/(1.5 \times 1.2 \times 1.0)$). This is the start of 'backcasting' – identifying a future desired state; in this case stabilising CO_2 emissions. Expressed in terms of average car fuel economy, this index figure would represent improving average UK car fuel economy from the current figure of 9.1 litres per 100 km to 5.1 litres per 100 km. Equivalent values for the motorist in terms of fuel consumption would be an impressive 55 mpg (46 miles per US gallon), which is much higher than the current average.

At the global level, the emissions index would need to be 0.28 to hold transport's CO_2 emissions at the 2005 baseline levels, requiring a fourfold improvement in fuel economy. So, within 20 years, the world's car fleet would need to average about 2.6 litres per 100 km. This very low fuel consumption would translate to approximately 110 miles per UK gallon (or 92 miles per US gallon).

1.4 Reducing transport's environmental impacts

The thought of achieving a global average car fuel economy of 2.6 litres per 100 km within 20 years suggests that the sums are starting to look beyond the realms of political (and possibly technical) viability. But this is without even attempting to *reduce* CO_2 emissions from personal transport. Successive reports of the Intergovernmental Panel on Climate Change (IPCC; for example, Houghton et al., 1990 and Watson et al., 2001) have suggested that a 60% cut on 1990 levels is needed to mitigate the effects of climate change. Following associated reports by the UK Royal Commission on Environmental Pollution (RCEP) advocating a 60% cut in UK CO_2 emissions by 2050, with a 40% cut by 2020, the UK government adopted a target to cut CO_2 emissions by 20% by 2010. In the 2003 Energy White Paper, the UK government announced a further long-term target to reduce UK CO_2 emissions by 60% by 2050.

Taking the 2020 target of a 40% cut, and arbitrarily easing it back to 2025, what sort of efficiency improvements might achieve a 40% drop in emissions? Returning to consider the UK situation, in 2005, transport's CO_2 emissions had already risen by over 15% since the IPCC's baseline date of 1990. To adopt 2005 as a baseline year requires adjusting the figures to allow for this growth. This can be done by making the target index for CO_2 emissions to be *not* 0.6, but 15% lower at 0.52. This becomes the target index figure to aim for. Again, if we were to rely on efficiency measures alone, the index for emissions per vehicle kilometre would need to be reduced to around 29% of current levels (Table 1.5).

Table 1.5 Vehicle Efficiency Improvement required to achieve a 40% reduction in CO_2 emissions by 2025

Population	×	**Car journeys per person**	×	**Journey length**	×	**Emissions per vehicle km**	=	**Total emissions**
1.0	×	1.5	×	1.2	×	0.29	=	**0.52**

So, if only fossil fuels were used, and every other factor follows the BAU forecast, then fuel economy would need to improve about threefold, to an average of 2.6 litres per 100 km. Allowing for a proportion of less fuel efficient vehicles, much of the car fleet would need to achieve under 2 litres per 100 km. Could such an improvement be achieved in 20 years?

The use of smaller-engined, more economical cars can cut fuel use and CO_2 emissions substantially. There are a number of car designs, such as the two-seater Smart and some smaller diesel models (see Figure 1.4), that achieve around 3 litres per 100 km. However, their use is not widespread. There is a very big difference between there being some cars that can get close to the target of a fuel economy of less than 2 litres per 100 km and the *whole* car fleet managing that within 20 years. Currently the UK car fleet has an average fuel economy of 9.1 litres per 100 km. This is not the best, or the worst. Italy, for example, has an average fuel economy of 7.5 litres per 100 km, which is associated with a vehicle tax regime strongly favouring smaller-engined cars. The USA, for example, averages 11.6 litres per 100 km (and there, the increasing use of 4-wheel drive sports utility vehicles is pushing up fuel consumption).

Figure 1.4 (a) The two-seater Smart car; (b) there have been a number of other designs like this electric micro car. However the Smart is the only one produced in serious numbers

Can car fuel economy be massively improved? In the UK, a study by Cousins and Sears (1997) explored what level of fuel economy could be achieved using best practice current technology. This project sought to produce not just a highly fuel-efficient car, but one that would win consumer acceptance. Their study selected a four-seat, five-door family hatchback powered by a 600 cc, 23 kw petrol engine, producing a top speed of 152 kph and a performance and price comparable to contemporary small cars (e.g. the Corsa 1.0), but with fuel consumption averaging 2.5 litres per 100 km.

A technology recently introduced to the market is the hybrid engine. By having both electric and internal combustion engines in one vehicle, each type can be utilised at high efficiency. The internal combustion engine is run more constantly, with the electric motor used in slow, stop-start conditions and when strong acceleration is needed. One of the first hybrid cars introduced to the European market, the Toyota Prius, has a test fuel consumption of 4.9 litres per 100 km.

These sorts of technology look as though could deliver an average fuel economy in the 3–5 litres per 100 km range, which is somewhat over halfway to the 20-year target of 2.6 litres per 100 km, if, of course, people were willing to accept such vehicles. Even though the above designs took performance and consumer acceptability into account, people have been very reluctant to buy fuel-efficient vehicles. Sales of the Prius and other hybrids are small, although the Smart has captured a reasonable 'second urban car' niche market. Yet to achieve a 2025 vision of a 3 or 4 litres per 100 km car fleet would require a substantial change in what we think of as a 'car'. It is a future where there would be very few large, heavy or high-performance cars. We would probably need to say goodbye to gas-guzzling 4 × 4 multi-purpose all-terrain vehicles, so beloved for use in the urban school run; there would be no room for them in such a high fuel economy future. The vast bulk of the car fleet would have to be modest, low-accelerating vehicles if such good fuel consumption were to be achieved in practice. A key lesson from this simple modelling exercise is that a seemingly 'technical fix' approach would require considerable behavioural change to make it work.

Of course, this ambitious level of fuel economy is referring only to developed nations. The necessary improvement in fuel economy becomes even greater once a global perspective is taken. Globally, if the population increases by 30%, consumption will increase and to reach the target of 0.52, emissions per vehicle km will have to reduce to an index of 0.14.

Table 1.6 Global vehicle efficiency improvement required to achieve CO_2 target

Population	×	Car journeys per person	×	Journey length	×	Emissions per vehicle km	=	Total emissions
1.3	×	2.3	×	1.2	×	0.14	=	**0.52**

If using conventional fossil fuels, this represents a global average fuel consumption of 1.3 litres per 100 km. Taking a longer perspective with further population and car use growth, this would need to improve even more.

1.5 Alternative fuels and renewable energy

For a developed country like the UK, it looks as if the widespread use of fuel-efficient vehicle technologies could get us about halfway to a CO_2 emissions reduction target. At a global level this approach looks far less hopeful, as fuel economy improvements, even if achievable, are set to be overwhelmed by a massive increase in consumption. Given such a trend, might the use of less-carbon-intensive 'alternative fuels' be the answer? Carbon intensity (the amount of carbon released in combustion per unit of energy generated) could be added to the index model. This would allow backcasting to explore the effect of a cut in carbon intensity. This is shown in Table 1.7, which splits 'Emissions per vehicle km' into two separate components. The first is 'Fuel per vehicle km', which is the fuel consumption of a vehicle, and the second is 'carbon intensity', which is how much CO_2 is within the fuel used. Table 1.7 calculates by how much the carbon intensity of fuels needs to drop if all other factors in the formula remain at the BAU rate we started with in Table 1.3. Therefore 'fuel per vehicle km' drops to 0.9. This shows that a threefold reduction in carbon intensity would be needed to hit the UK target of a 40% reduction by 2025.

Table 1.7 UK population and journey number reduction, and vehicle efficiency improvement required to achieve CO_2 Target in 2025, baselined at 2005

Population	×	Car journeys per person	×	Journey length	×	Fuel per vehicle km	×	Carbon Intensity	=	Total emissions
1.0	×	1.5	×	1.2	×	0.9	×	0.32	=	**0.52**

Is this possible in 20 years – even if people were willing to accept the type of vehicles concerned? This reduction in carbon intensity is beyond what most alternative fuels can offer. Chapter 2 of this book will look at alternative vehicle fuels and engines in some detail. Here the emphasis is on the overall environmental performance possible from such fuel changes and on exploring the scale of changes needed to achieve such an improvement. A review of studies of the carbon intensity of alternative transport fuels (detailed in Potter and Warren, 2006, pp. 68–70) shows that most alternative fuels have been developed to reduce local air pollutants (such as the notorious Los Angeles smog), with little consideration for CO_2 emissions. The first approach is simply to compare petrol and diesel, as diesel is an already widespread alternative to petrol. The CO_2 equivalent life cycle emissions from diesel are about 25% lower than from petrol-engined vehicles. Compressed natural gas (CNG) has been in widespread use as a road vehicle fuel in some countries (e.g. Italy and New Zealand) for many years and is emerging as the leading alternative fuel in a number of European states, because it enables significant air quality improvements to be achieved. It is also a relatively simple matter to adapt existing engines and designs to use it. However, although CNG produces considerably lower air pollutant levels than petrol or diesel, in terms of climate change gas emissions the reduction is marginal. CO_2 from CNG vehicles is only about 15% lower than for petrol vehicles.

If electrically powered vehicles are used, the effect on CO_2 emissions depends on the primary energy source used to generate the electricity and the efficiency of the generation process. For the average European mix of electricity-generating fuels, an electric car achieves about a 40% cut in the emission of greenhouse gases compared with a petrol car, and a 22% cut compared with diesel. However, if coal is the source of primary energy, the CO_2 emissions are slightly worse than those from a petrol-engined car. If gas is used for electricity generation, greenhouse emissions from electric

Figure 1.5 A vehicle powered by liquefied petroleum gas (LPG). Like CNG, LPG cuts local air pollution but CO_2 emissions are only marginally less than for conventional road fuels

vehicles drop by 50% compared to petrol, while electricity from nuclear and hydroelectric power stations produce the greatest improvement (by 85%). Although nuclear and most renewable energy sources produce little or no CO_2 during electricity generation, CO_2 is produced in building and maintaining the power stations, which is reflected in these fuel life cycle figures.

The use of biofuels was emphasised in the 2003 Energy White Paper (DTI, 2003). This anticipated that by 2020 up to 5% of transport fuels could be bio-diesel and bio-ethanol. However, biofuels are very mixed in their effect on CO_2 emissions (Figure 1.6). If a car uses an ethanol fuel produced from maize and other crops that are energy intensive to grow and where the fuel manufacturing process is also energy intensive, there is little or no improvement over fossil fuels (DTI, 2000). This is particularly so where fossil fuel is used in the distillation process for maize. However, ethanol from wood comes out well, offering a 66% cut in CO_2 emissions as the production of this fuel is less energy intensive. This is also true of bio-diesel, produced from rape seed.

The use of hydrogen fuel cells for automotive applications has attracted a lot of attention, and they look like replacing the battery-electric vehicles as the main challenger to the internal combustion engine (Chapter 2 covers this issue in some detail). Fuel cells produce electricity using an electrolytic process, converting hydrogen and oxygen to electricity and heat, with water vapour the only emission. But with hydrogen, as with any other manufactured fuel, it is important to take into account the carbon content of any primary fuel used to manufacture it, and other overall life cycle emissions. Hydrogen produced from renewable energy sources would be a very clean fuel in terms of both local air pollutants and CO_2 emissions. This is the assumption in Figure 1.6, whereas if hydrogen were produced (as at present) from oil, the CO_2 emissions would be similar to those for LPG. At the moment (and for a good while to come), electricity produced from renewable sources seems likely to be used for existing domestic and commercial purposes. It seems unlikely that in a 20-year timeframe there will be enough renewable energy capacity for both existing electricity markets and a new market for hydrogen used in transport.

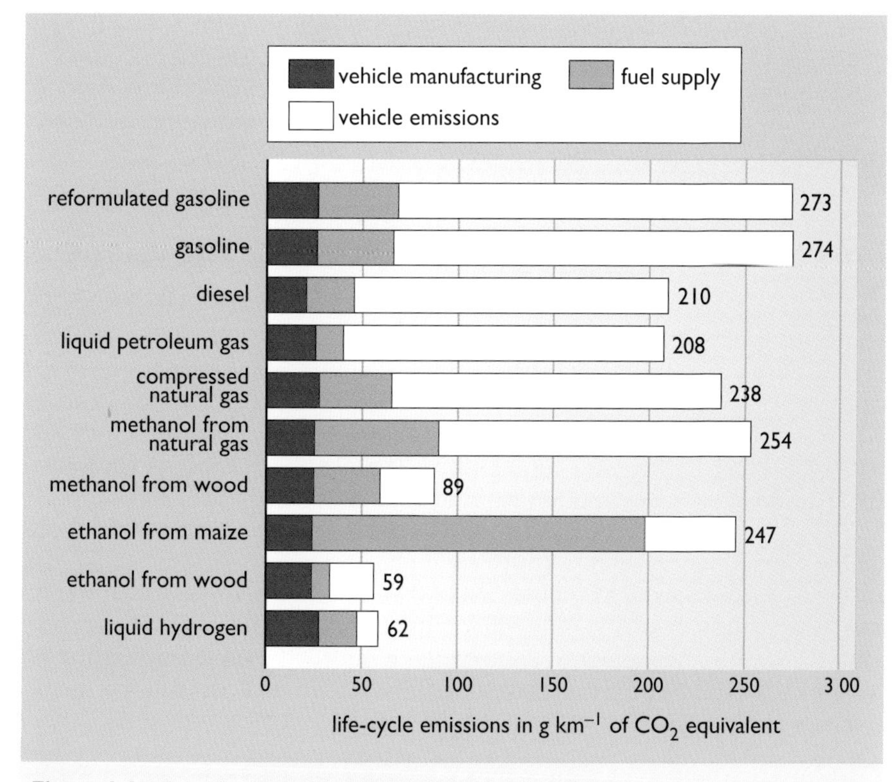

Figure 1.6 Life cycle climate change gas emissions from alternative fuels (source: OECD, 1993)

This issue was explored in the report, *Fuelling Road Transport: Implications for Energy* (Eyre, Fergusson and Mills, 2002). This concluded that until there is a surplus of electricity generated from renewable sources, it is not beneficial in terms of carbon reduction to use renewable electricity to produce hydrogen for any application, transport or otherwise. The report continues:

> Higher carbon savings will be achieved through displacing electricity from fossil fuel power stations. There would be some carbon savings from hydrogen vehicles using electricity from a power station dependent largely on gas and renewables, if the gas technologies are combined heat and power (CHP). But the supply of hydrogen to mass-market vehicle applications is likely to require more electricity than can be supplied from renewables and CHP alone for at least 30 years.
>
> Eyre, Fergusson and Mills, 2002, p. 4

The situation appears to be that areas other than transport should be prioritised for the use of renewable energy. This report further concludes that the cheapest route to hydrogen would be to produce it from natural gas and that this has 'some potential carbon benefits if used in high efficiency fuel cells vehicles', but that the benefits are relatively small compared to diesel and petrol hybrid vehicles. Overall, the use of alternative fuels for transport presents a very mixed, and extremely uncertain, picture. Over the next 20 years fuel cells seem set to become a mainstream automotive technology, but in terms of CO_2 emissions there may be anything from little effect to a 60% improvement under very optimistic assumptions. Even the latter does not quite make the 68% improvement suggested above as necessary if other consumption factors were to remain unchanged. In the longer term (30 years or more), hydrogen generated from renewable energy sources would provide an ultimate answer, but we are a long way from having sufficient supplies of renewably generated electricity to achieve this.

Fuel efficiency and alternative fuels – conclusions

This analysis of the role of fuel efficiency and switching to alternative fuels leads to an important conclusion. Even if a purely technical fix, product-level approach were taken, only a combined strategy of *both* improving fuel economy *and* developing alternative fuels stands any hope of getting CO_2 emissions down to a sustainable level. For example, a doubling of fuel economy (to an index value of 0.5) plus a halving of the carbon content of fuel (to an index value of 0.55) would hit the UK target. This is potentially achievable, but would require the use of only the very best alternative fuels. At the global level even this approach looks hazardous. Because of the rise in consumption, even if fuel economy doubled, carbon intensity would have to be cut by nearly 80%, which implies that virtually all primary fuel would have to be nuclear or renewable. Even if fuel economy improved by a factor of about four, carbon intensity would still need to be nearly halved (Table 1.8.).

Table 1.8 Efficiency and fuel improvement to achieve global CO_2 target

Population	×	Car journeys per person	×	Journey length	×	Emissions per vehicle km	×	Carbon Intensity	=	Total emissions
1.3	×	2.3	×	1.2	×	0.25	×	0.58	=	**0.52**

This backcasting scenario is one of highly efficiently produced hydrogen (or other cleaner fuels) powering rather small vehicles. These vehicles would need to have an extremely good fuel economy. The index figure of 0.25 for 'fuel per vehicle km' represents the petrol equivalent of 2.3 litres per 100 km compared to 9.1 today. This seems pretty unlikely to be achieved in 20 years. It is hard to envisage that this could be achieved politically, even though it may be just about technically possible. Added to this, beyond 2025, further cuts in CO_2 are required. The IPCC target is for an eventual cut of at least 60% in global emissions compared to 1990, which the UK government set as a long-term domestic target in the 2003 Energy White Paper. To compound the problem, car use and the population will inevitably rise and so overall consumption will increase, counterbalancing any individual vehicle improvements. As time passes the goalposts move and the whole situation becomes even more challenging.

A 60% cut in CO_2 compared to 1990 would require the index figure for total global emissions to be cut to a value of 0.34. It is hard to speculate with any accuracy how the key factors in the index model will have changed beyond 2025, as it is simply so far ahead. You can put your estimates into the index model and work out what the carbon intensity figure needs to be to hit the 0.34 target for total emissions. My own workings suggest carbon intensity would need to drop to 0.15 or less. What does such a figure mean?

One way this index figure could be achieved is if each car uses a quarter of the energy of those around today and gets only an average of 15% of its energy from fossil fuels. The other 85% would have to come from renewable sources. The overall result (allowing for the higher energy efficiency as well) is that the average car in 2050 will need to run on under 4% of the fossil fuels used by the average car today.

Looking this far ahead it could be argued that by then a global, totally carbon-free energy supply system will have evolved. However this exercise does show the magnitude of the challenge ahead. Returning to our more comprehensible 20-year timescale, a very strong technical fix approach might just about achieve the 20-year target at the level of a developed economy like that of the UK. At the global level, the necessary improvements in fuel economy and type of fuel, even for a 20-year timescale, look unrealistic. Once a longer timescale is envisaged, the whole situation is far more uncertain.

It is also important to add to this that our analysis has been looking only at emissions and energy use arising from transport activities. As was briefly mentioned at the beginning of this chapter, there are a number of important transport policy issues that would be unaffected by using technical measures to reduce emissions. These include traffic accidents, traffic congestion and the host of health-related issues linked to sedentary, car-oriented lifestyles. All these issues are about the amount of motor traffic, rather than how it is powered.

1.6 Travel mode and volume of travel

The magnitude of the changes required in only 20 years looks daunting when a purely 'technical fix' approach to cutting emissions from transport is taken. So, can changes to the 'consumption' elements in our simple index model suggest a more viable path? This would also mean that other issues relating to the volume of traffic would be addressed, such as congestion, accidents and adverse health effects.

In the index model, consumption aspects are expressed in terms of the number and length of journeys. A much-advocated approach is to cut transport's environmental impacts by somehow shifting trips ('modal shift') from the car to less energy-intensive forms of transport. To evaluate this option requires a return to a UK focus, as it is difficult to obtain and use global figures for key factors such as the share of travel by each mode of transport (modal share) and journey length.

A number of studies (detailed in Potter, 2003) have compiled empirical information on the quantities of energy and CO_2 emissions arising from the operations of various transport modes. Table 1.9 is a compilation from these sources for a range of urban public transport vehicles. These figures cover the entire fuel life cycle, allowing for the different engine efficiencies and fuel-production systems and the differing carbon contents of the fuels concerned. Clearly, the energy use figures quoted depend very much upon the individual design of vehicles and where and how they operate. The above studies (Potter, 2003) do note variations. The information for buses was provided by a number of UK urban bus companies and that for railways by London suburban rail operators. The light rail figures were provided for the modern tram operations in Manchester and the metro/underground figure is for the London Underground.

The data have been compared with other UK and European studies of energy and CO_2 emissions (CEC, 1992; Best Foot Forward, 2000; Climate Care, 2000). This comparison suggests that the energy use and CO_2 emission figures for buses and diesel trains are broadly similar to those found in other developed countries. For electric trains, light rail and metros, the energy-use figures are also broadly similar to the UK figures, but CO_2 emissions will vary according to the primary fuel mix of the power stations. The 2000 UK mix of gas, coal and nuclear generation was estimated to produce 480 grams of CO_2 per kilowatt hour (kWh) of electricity generated (modern coal power stations produce about 950 grams of CO_2 per kWh of electricity generated and gas combined cycle stations about 450 grams (Eyre, 1990). The UK average figure also includes oil, hydro and nuclear generation).

The first three columns of Table 1.9 contain the information for the size of vehicles and the CO_2 per vehicle km. The next column simply divides the vehicle CO_2 emissions by number of seats to produce grams of CO_2 per seat km. This shows, for example, that a full electric train has half the emissions of a car with all five seats occupied. In general, the slower forms of public transport produce the least CO_2 emissions.

Table 1.9 Fuel life cycle CO_2 emissions for major transport modes

Mode	Seats	Kg CO_2 per vehicle km	Grams CO_2 per seat km	Grams CO_2 per person km – peak travel	Grams CO_2 per person km – off-peak travel
Urban electric train	300	11.7	39	65	156
Urban diesel train	146	8.8	60	98	240
Light rail	265	10.1	38	54	95
Metro/underground	555	26.0	46	66	115
Single-deck bus	49	1.6	33	66	165
Double-deck bus	74	1.9	26	52	130
Minibus	20	0.8	40	57	200
Medium-sized car	5	0.39	78	339	195

Source: Potter, 2003

In practice, all seats are not occupied and so the final two columns contain estimates for how well occupied vehicles are for peak and off-peak travel. For example, actual peak-hour car occupancy in Britain averages only 1.17 persons, while trains and buses are near fully loaded. In the off-peak, the situation is different. For shopping, leisure and holiday trips, car occupancy is in the range of two to three persons (50–60%) and off-peak loadings of public transport average 40% or less. This is important, as the relative performance of car compared to public transport will vary by whether it is peak or off-peak trips that are involved. In the peak hour, CO_2 emissions from cars per person kilometre is over five times that of someone travelling in an electric train; for off-peak travel, a well loaded car actually produces slightly less CO_2 than a poorly loaded electric train.

To explore the effect of modal shift requires the formula model to be split into the three main components of motorised travel: car, bus and rail. This is not to say that non-motorised travel (walk and cycle) is unimportant, but it does not generate CO_2 to any significant extent. Trip shifting to walk and cycle can be accommodated in the model by cutting the 'journeys per person' figure for the motorised modes. In the UK, cars account for 88% of motorised trips, buses 10% and trains 2%.

This reworking of the baseline index is shown in Table 1.10; it is taken that the energy use per passenger kilometre by train and bus is, on average, about half that of cars. This ratio is based upon the information discussed above. It could be argued that in some circumstances the energy efficiency of public transport may be somewhat better, but a halving is viewed as a safe estimate. The carbon intensity is similar for all three as oil is the main fuel used for all transport.

Table 1.10 Expanded UK baseline CO_2 emissions index for all transport modes: car, bus and rail (2005)

	Journeys per person	×	Journey length	×	Energy use per person km	×	Carbon intensity	×	Modal share		
Car	1.0	×	1.0	×	1.1	×	1.0	×	0.88	=	0.97
Bus	1.0	×	1.0	×	0.5	×	1.0	×	0.10	=	0.05
Rail	1.0	×	1.0	×	0.6	×	1.0	×	0.02	=	0.01
Total emissions for all modes										=	**1.03***

*Not exactly 1.0, but can be rounded down
Source: DfT, 2004a

A backcasting modal shift scenario could be based around the targets suggested by the UK's Royal Commission on Environmental Pollution (1994), which have been used widely in transport policy development. How might these targets be achieved? Some practical examples of measures will be considered in Chapters 3 and 4. Pricing measures, although far from popular, are effective. The introduction in February 2003 of a £5 congestion charge to drive within Central London cut traffic levels by over 20%. London is the latest of a number of cities to introduce such a scheme. The first was Singapore in 1975, where traffic levels were reduced in a similar way to London and, with regular adjustments to the charging system, road traffic has been held at that lower level ever since. Such pricing schemes often require technological innovation as well. The Central London Congestion Charging Scheme, for example, is operated through a network of number plate recognition cameras that can distinguish (by links to computer databases) between motorists who have, and those who have not, paid the charge and also identifies those who are exempt. The system can also allow for temporary exemptions, such as an accident leading to traffic being diverted through the congestion charging zone, by ensuring those vehicles are not fined for non-payment. Exemptions to London's congestion charge, incidentally, include alternative fuel cars, so this behavioural change mechanism also stimulates a technological response as well (sales of hybrid cars in London are relatively high (Harrison, 2005)).

Figure 1.7 The London Congestion Charging Zone

In the following version of the index model it is assumed that, over 20 years, pricing and a whole variety of other modal shift measures will have cut the car's share from 88% of motorised trips to 65%, with the bus share increasing to 25% and rail's to 10%. In order to show what this can do alone, no technical improvement measures are included. Thus changes to fuel economy are at the BAU rate, with the index for energy use per passenger km improving from 1.1 to 0.97 for car, and improving to 0.44 for bus and 0.53 for rail. It is assumed that there will be a continuing use of oil-based fuels, so there will be no change in carbon intensity (the index figure remaining at 1.0 for all modes).

Table 1.11 UK potential scenario modal shift and CO_2 emissions in 2025

	Journeys per person		Journey length		Energy use per person km		Carbon intensity		Modal share		Total emissions
Car	1.5	×	1.2	×	0.97	×	1.0	×	0.65	=	1.13
Bus	1.5	×	1.2	×	0.44	×	1.0	×	0.25	=	0.20
Rail	1.5	×	1.2	×	0.53	×	1.0	×	0.10	=	0.09
Total emissions for all modes										=	**1.42**

The net result, surprisingly, is a 42% increase in CO_2 emissions. This may be better than the business as usual, 60% rise in CO_2 without modal shift, as considered earlier (Table 1.3), but the cut in CO_2 arising from modal shift is insufficient to counterbalance the rise in other behavioural factors in the model. An important component of this is trip lengthening, which involves not only motorised trips becoming longer, but also the substitution of short trips on foot with longer trips by car, which is reflected in the rise in the number of journeys per person. Simply to get the total emissions in the index model to equal

1.0 would require the very unlikely combination of the car modal share being cut to 30%, with the bus share rising to 40% and the rail share to 30%. Even this would only hold CO_2 emissions at their current unsustainable level.

This simple exercise leads to an important conclusion. Not only will the technical fix not work in isolation, but modal shift, as an isolated policy, is also doomed to failure as a CO_2 reduction measure.

1.7 A multiple approach

This backcasting exercise using a simple index model demonstrates clearly that the only technically (and certainly politically) practical way in which transport's CO_2 emissions can be cut to sustainable levels is to combine changes in *both* the vehicle technology (fuel efficiency and fuel type) and *all* behavioural factors. This is called the 'multiple' approach. Importantly, behavioural change cannot just involve modal shift between different forms of motorised transport. Behavioural change needs to involve a reduction in the trend of increasing trip lengths and the effect this has on non-motorised travel. A major factor in the increase in road traffic in recent years has been because we make longer trips.

One variation of the index formula that would achieve a total emissions index figure of 0.52, is shown in Table 1.12.

Table 1.12 A multiple approach to achieve UK CO_2 target in 2025

	Journeys per person		Journey length		Energy use per passenger km		Carbon intensity		Modal share		Total
Car	1.3	×	1.1	×	0.55	×	0.8	×	0.65	=	0.41
Bus	1.3	×	1.1	×	0.25	×	0.8	×	0.25	=	0.07
Rail	1.5	×	1.2	×	0.30	×	0.7	×	0.10	=	0.04
Total emissions for all modes										=	**0.52**

This particular combination involves:

- a 30% increase in car and bus journeys (rather than 50% in the BAU scenario);
- halving the increase in trip length for car and bus (from 20% to 10%); rail is at 20% assuming this picks up some long car trips that transfer to rail;
- a 50% improvement in energy use per person km for all modes. This could be a combination of better vehicle fuel economy and increased vehicle occupancy;
- a 20% cut in the carbon intensity of the fuel used for road vehicles and a 30% cut for rail (the latter probably through electrification and/or bio-diesel trains);
- modal shift as in the RCEP report, cutting car from 88% to 65% of motorised trips, with bus rising to 25% and train to 10%.

The first two factors in this list would involve the proportion of walking and cycling trips being retained or increased, through the use of land use planning policies that reduce the need for motorised travel (through higher densities and fewer car-based, out-of-town developments).

The multiple approach – conclusions

The 50% improvement in fuel economy is reflected in the index figures in the column for energy use per person km. This could be achieved by a combination of better fuel economy and also increased vehicle occupancy. Overall, for the UK, the scenario outlined above means an improvement from our current average car fuel economy of 9.1 litres per 100 km to the equivalent of 4.5 litres per 100 km (approximately 63 mpg (UK) or 53 mpg (USA)), which is a tough 20-year target, but is probably both technically and politically possible. This improvement in fuel efficiency needs to be combined with the development of alternative fuels to occupy about a third of the market. This also appears to be a tough, but reasonable 20-year aspiration. There would also have to be significant modal shift and a reduction in the rate of journey length increase to hit the CO_2 reduction target recommended by the scientific community. The number and length of journeys are crucial factors, and yet these are rarely considered in the transport/environment debate.

If all travel generation factors are not addressed, an unrealistic improvement in individual factors is required, as we have explored when looking at technical fix and modal shift options in isolation.

1.8 Reducing transport dependency

The need to reduce the number and length of trips, plus the need to reduce motorised travel, brings in the crucial issue of 'intelligent consumption' with transport systems that deliver the functions of mobility at a lower energy and resource cost. If access to people, facilities and goods can be achieved with less mobility, then this could make an important contribution to cutting transport's environmental impacts. Reducing the need to travel frequently leads to a discussion of land use planning policies and the need to increase urban densities in order to cut the need to travel. However, this is but one part of reducing transport dependence. Indeed any policy that simply relies on packing people so close together that traffic congestion eventually cuts car use, is probably as doomed to failure as any other single policy measure. There is limited experience of how to travel differently and to enhance accessibility. Some technologies and alternative systems have a potential to reinvent accessibility and mobility in ways that can cut environmental impacts. Again, a multiple approach of complementary measures seems appropriate, but this is very much an area of uncertainty, where further understanding is desperately needed.

IT and travel substitution/generation

The travel-substituting potential of the internet revolution appears, on the face of it, to be strong. I am, at the moment, writing this chapter from home, where I can email colleagues, send and receive documents and have access to a full library and the complete (and overwhelmingly distracting) information resources of the World Wide Web. All this comes to me in my study at the back of my house. In consequence, I go into the Open University's campus only two or three times a week. The travel reduction potential of such 'telecommuting' seems obvious; or is it?

A survey of Californian telecommuters by Pendyala et al. (1991) provides strong evidence of the positive transport effects of telecommuting. Not only was car use for commuting purposes cut, but non-work trips were also reduced. It appears that once telecommuters no longer have a long drive to work, driving long distances for other purposes becomes less acceptable and they tend to undertake shopping and leisure trips more locally.

But there are negative as well as positive 'rebound' effects with this seemingly beneficial technology. Firstly, if telecommuting results in increased energy use in the home, particularly for heating and air-conditioning, then the overall energy and environmental improvements will be less than envisaged. This again reinforces the need for a life cycle and systems analysis. Of possibly more significance are longer-term lifestyle adjustments to a communications-intensive society. Historically, improvements in the availability and speed of travel have not led people to travel less than they did before, but have always led to lifestyle changes that have resulted in more motorised travel being generated. So, for example, the arrival of buses and trams did not result in people getting to work faster: they moved further away from work and created suburbs. If people need to travel to a place of work on only two or three days a week (or fewer), this is likely to lead them to live further from work, possibly in more remote locations that are very car dependent for all travel needs (Potter, 1997). In an in-depth analysis of the implications of the 'information society' for spatial planning, Graham and Marvin noted that:

> Rather than simply being replaced, transport demands at all scales are rising in parallel with exploding use of telecommunications. Both feed off each other in complex ways, and the shift is towards a highly mobile and communications-intensive society.
>
> Graham and Marvin, 1996

There is a real danger that IT and telecommuting could well result in the generation of more travel than they eliminate.

Reinventing car 'ownership'

An alternative way to manage the use of the car involves not physical or electronic controls, but reinventing the way we obtain and pay for car use. This brings us back to the issue of pricing and economics, which cannot be avoided in transport policy studies. To buy a car involves high fixed costs and once this is made the only relevant costs are those for running it. The most common perception is that fuel is the only cost of a car journey. For public transport, the cost structure is different. There are no separate 'capital' or 'running' costs; all the costs are combined into the price of a ticket. This different way of paying for travel stacks the odds against the bus and train, and, with most of car costs being fixed, there is also little disincentive when trip lengthening occurs. When making an individual journey, it is usual for a car user to compare the fuel costs of travelling by car with the price of a bus or rail ticket. If paying for car use were different, and (as for public transport) the capital and other fixed costs were included in a 'pay by the kilometre' charge, then it is likely that perceptions of the relative cost of car and public transport, and also of short as opposed to longer car journeys, would be different.

One example of this is the Car Club concept, in which a fleet of cars is available to individuals who pay for all costs by the kilometre (see Figure 1.8). Car-sharing clubs are most widespread in Switzerland, the Netherlands and in parts of Germany, where studies have shown how they affect people's travel patterns. In Switzerland, Harms and Truffer (1999) concluded that car sharing reduced the distances travelled by car. Their research looked at both former car owners who joined the Swiss Mobility Car Club and former non-car owners. Before they joined the car club, former car owners drove less than average. They covered about 9300 km per year by car compared with a Swiss average of 13 000 km. This is to be expected, as people who did not drive much might feel that the fixed cost of car ownership was a lot compared with their limited use of a car. These people would be most attracted by a car club scheme. Even though these people already drove relatively little by Swiss standards, after they became a car-sharing member this was reduced to 2600 km per year, which is 28% of the distance they previously drove.

Figure 1.8 A car and user of the Swiss Mobility Car Club. Half the Swiss population lives within a 10-minute walk of a Mobility car park. The inset shows the screen-mounted smart card reader that gives the user access to the car they have booked

Some of the reduction in car travel for the car club members involved a shift to public transport, bike or motorbike. This accounted for 4000 km, which is about 60% of the reduction in car use. Significantly, the other 40% was produced by people cutting trip lengths or finding another way to do things that did not involve travelling at all. Surprisingly, former households without cars did not drive more after they joined the car club. It appears that most of them already had some access to borrowing or hiring cars and the car club was simply a better or cheaper way to carry on doing this.

Harms and Truffer emphasise that the changes in mobility patterns should not be totally attributed to the car-sharing system. Joining a car club was sometimes associated with other changes in people's lives, such as moving to another town with different conditions for private and public transport, getting a new job in a different place or with different working conditions, or changes in income. Transport factors did play a certain role. In some cases joining the car club was triggered by a terminal breakdown of their own car or by increasing difficulties with parking, congestion or repair costs of their car.

Overall, maturing car club schemes do suggest that changing the way cars are paid for can have a significant effect upon the mode of travel used, the distance people travel and whether travelling is seen as necessary at all. Forms of obtaining access to cars like car clubs could be developed, particularly if the taxation system were to favour them. However, the tax system could also produce a 'car club' effect even for continued private car ownership. Ubbels, Rietveld and Peeters (2002) explored the impacts on car use and the environment of replacing existing taxation on cars and fuel with a kilometre charge for using roads. The redistribution of fixed taxes to a kilometre charge resulted in a modelled reduction in car kilometres travelled of between 18% and 35% compared to the base case. CO_2 emissions from cars were cut by 22–40% and NO_x by 40–50%. Total travel declined by only 5–10%, but interestingly public transport travel increased by only a maximum of 5%. The main impact of the kilometre charge was to increase walking and cycle use by 5–10% and to increase car occupancy.

In 2004, the UK government announced plans for a national road pricing scheme to be introduced within 10–15 years (DfT, 2004b), which will partly or wholly replace Fuel Duty and Vehicle Excise Duty ('car tax'). All cars would be instrumented with a GPS system, with there being a mileage charge that would vary according to whether motorists drove on congested roads, at peak times, or on relatively clear roads in off-peak periods. Oregon State in the USA is also planning to introduce a distance charge to replace fuel duty, and distance charging systems are already used for heavy goods vehicles in Germany and Switzerland. The UK proposals have been criticised for not varying the charge by the fuel economy or environmental performance of the vehicle (Potter and Parkhurst, 2005), but it is clear that a number of countries are now moving to a 'pay as you drive' tax system that will have significant effects on the cost structure of motoring that could change the number and length of trips.

The effects of taxation and institutional changes, such as the ways in which car use is obtained and paid for, would be reflected in the part of our index model relating to the number of trips and their length. It would also result in some modal shift effects as well.

1.9 Conclusions: travelling lightly

This chapter has explored a framework for thinking through how the personal transport sector could achieve a sustainable level of CO_2 emissions that would meet climate change targets in the medium to long term. It has shown that it is necessary to address all factors generating the overall volume and emissions from the personal transport sector.

These include:

- fuel efficiency of vehicle involved
- carbon content of fuels used
- number of journeys made
- journey length
- vehicle occupation
- mode of transport.

Although this backcasting model is a simple one, the overall results are consistent with more sophisticated backcasting exercises. For example, the conclusions of a study conducted by the OECD (Potter et al., 1998) were that to cut all environmentally damaging emissions from transport to sustainable levels would require a third of the reduction to come from technical measures and two-thirds from demand management. Hickman and Banister's backcasting study for the UK Department for Transport (Hickman and Banister, 2006) comes to a remarkably similar conclusion to our simplified exercise, suggesting radical improvements in car fuel economy and low-carbon fuels are needed, together with European best-practice levels of walking, cycling and public transport use becoming the norm.

Such a multiple approach to reducing transport's environmental impacts requires a good understanding of how these factors interact as a system. The rest of this book explores this issue, starting by examining potential technical improvements to vehicles and then moving on to behavioural change, with a particular emphasis on how policy measures can be practically applied by organisations whose activities generate substantial travel needs.

Despite studies showing the necessity for a multiple approach to provide any hope of achieving transport sustainability, technical measures have come to be viewed as politically easier to promote but their limitations are poorly understood, or simply not considered. Behavioural consumption policies are much talked of, but are rarely applied to an effective extent and generally fail to address the full range of consumption factors involved. In particular, approaches to reduce transport dependence appear to offer much potential, but are rarely considered.

Added to all this, there is a serious issue of the differences in timing between the technical fix approach and the 'intelligent consumption', behavioural change approach. Some technical fix measures yield results more quickly than policies to affect change in travel behaviour. Thus a sensible approach would be to use the time that technical product and fuel change improvements can buy to put in place the longer-term 'intelligent consumption' behavioural change policies that will 'kick in' as the vehicle and fuel improvement effects start to wane. The political danger is that technical fixes, being seen as less politically sensitive, will be used to continually put off taking serious action to change behaviour until it is too late. The conclusion to this book will return to this and other issues around developing packages of policies to achieve sustainable transport systems.

References

Best Foot Forward (2000) http://www.bestfootforward.com [accessed 10 August 2006].

Bishop, S. and Grayling, T. (2003) 'The sky's the limit: policies for sustainable aviation', London, Institute for Public Policy Research.

Climate Care (2000) http://www.co2.org. [accessed 10 August 2006].

Commission for the European Communities (CEC) (1992) *Green Paper: The Impact of Transport on the Environment: a Community Strategy for 'Sustainable Mobility'*, Luxembourg, CEC Office for Official Publications.

Cousins, S. and Sears, K. (1997) 'E-Auto: the design of a 2.5l/100 km (113 mpg) environmental car using known technology', *Automobile Environmental Impact and Safety*, ImechE, pp. 369–79.

Department of Trade and Industry (DTI) (Annual) 'Digest of UK Energy Statistics', DTI, London, The Stationery Office.

Department of Trade and Industry (DTI) (2000) *The Report of the Alternative Fuels Group of the Cleaner Vehicle Task Force Report*, DTI, Automotive Directorate, London, The Stationery Office.

Department of Trade and Industry (DTI) (2002) *Energy consumption in the UK*, DTI [online] http://www.dti.gov.uk/energy [accessed June 2006].

Department of Trade and Industry (DTI) (2003) *Energy White Paper*, London, The Stationery Office (also at http://www.dti.gov.uk/energy/policy-strategy/energy-white-paper/page21223.html) [accessed 10 August 2006].

Department for Transport (DfT) (2004a) *Transport Statistics Great Britain*, London, The Stationery Office (Annual).

Department for Transport (DfT) (2004b) *Feasibility Study of Road Pricing in the UK*, London, DTI, The Stationary Office.

Ehrlich, P. and Ehrlich, A. (1990) *The Population Explosion*, New York, Simon and Schuster.

Ekins, P. et al. (1992) *Wealth Beyond all Measure: An Atlas of New Economics*, London, Gaia.

Eyre, N., Fergusson, M. and Mills, R. (2002) *Fuelling Road Transport: Implications for Energy Policy*, London, Energy Savings Trust.

Eyre, N.J., (1990) *Gaseous Emissions due to Electricity Fuel Cycles in the UK*, ETSU, Oxon, Harwell.

Graham, S. and Marvin, S. (1996) *Telecommunications and the City: Electronic Spaces, Urban Places*, London, Routledge, p. 296.

Harms, S. and Truffer, B. (1999) 'Car sharing as a socio-technical learning system: emergence and development of the Swiss car-sharing organisation', Mobility, http://www.ecoplan.org [accessed 10 August 2006].

Harrison, M. (2005) 'Jump in congestion charge to aid UK sales of Toyota's hybrid car', *The Independent*, 7 July.

Herring, H. (1999) 'Does energy efficiency save energy? The debate and its consequences', *Applied Energy*, 63, pp. 209–26.

Hickman, R. and Banister, D. (2006) 'Looking over the horizon', *Town and Country Planning*, vol. 75, no. 5, May, pp. 151–2.

Houghton, J.Y. et al. (eds) (1990) *Climate Change*, Cambridge, Cambridge University Press.

Hughes, P. (1993) *Personal Transport and the Greenhouse Effect*, London, Earthscan.

Mildenberger, U. and Khare, A. (2000) 'Planning for an environment friendly car', *Technovation*, vol. 20, no. 4, pp. 205–14.

Noble, B. and Potter, S. (1998) 'Travel patterns and journey purpose', *Transport Trends*, London, Department of Transport, Environment and the Regions, no. 1, pp. 3–14.

Organisation for Economic Cooperation and Development (OECD) (1993) 'Cars and climate change' Paris, OECD.

Pendyala, R., Goulias, K. and Kitamura, R. (1991) *Impact of Telecommuting on Spatial and Temporal Patterns of Household Travel: An Assessment for the State of California*, Davis, Institute of Transportation Studies, University of California.

Potter, S. (1997) 'Telematics and transport policy: making the connection', in Droege, P. (ed.) *Intelligent Cities*, Amsterdam, Elsevier.

Potter, S. (1998) 'Achieving a factor 10 improvement', in Daleus, L. and Schwartz, B. (eds), *Fšretag I Kretslopp*, Stockholm, Swedish Energy Agency, pp. 219–26.

Potter, S. (2003) Transport Energy and Emissions: Urban Public Transport, in Hensher, D. and Button, K. (eds) *Handbook in Transport 4: Transport and the Environment*, Amsterdam, Pergamon/Elsevier.

Potter, S. and Parkhurst, G. (2005) 'Transport policy and transport tax reform', *Public Money and Management*, vol. 25, no. 3, June, pp. 171–8.

Potter, S. and Warren, J. (2006) 'Travelling Light', Theme 2 of *T172 Working with our environment: technology for a sustainable future*, Milton Keynes, The Open University.

Potter, S., Enoch, M. and Fergusson, M. (2001) *Fuel Taxes and Beyond: UK Transport and Climate Change*, London, World Wide Fund and Transport 2000.

Potter, S., Lane, B. and Skinner, M. (1998) *G-8 Environment and Transport Futures Forum Proceedings*, EPA 160-R-98-002, Washington DC, EPA.

Royal Commission on Environmental Pollution (1994) *Transport and the Environment*, London, HMSO, 1994.

Teufel, D. et al. (1993) *Oko-Billanzen von Fahrzeugen*, Heidelberg, Umwelt und Prognose Institut.

The Economist (2005) 'Cars in China: dream machines', *The Economist*, 2 June [online] http://www.economist.com/business/displaystory.cfm?story_id=4032842 [accessed May 2006].

Ubbels, B., Rietveld, P. and Peeters, P. (2002) 'Environmental effects of a kilometre charge in road transport: an investigation for the Netherlands', *Transportation Research Part D*, vol. 7, no. 4, pp. 255–64.

Watson, R.T. and the Core Writing Team (eds) (2001) *Climate Change 2001: Synthesis Report*, A contribution of Working Groups I, II and III to the Third Assessment Report of the Intergovernmental Panel on Climate Change (IPCC), Cambridge University Press.

Chapter 2

Sustainable road transport technologies

by Ben Lane and James Warren

2.1 Introduction

There is no doubt that the invention of the internal combustion engine, along with the extraction of petrochemicals to produce refined motor fuel, has significantly shaped society and the natural environment over the last 100 years. In particular, the advent of motorised road transport, almost totally dependent on fuels derived from crude oil and on the internal combustion engine (ICE), has transformed most aspects of modern life.

Although the history of the automobile industry is well documented, in this section we focus on the main technological developments which have occurred in the automobile itself and some of the effects these have had on our environment. Once the technical obstacles had been overcome in the design of the 'Silent Otto' engine, it was the development of the moving production line by Ford in 1913 that first made the motor car widely available. Within ten years, Ford was selling over a million cars per year, and in some parts of the USA, car ownership reached one person in three, a ratio only reached in the UK in the 1970s. Figure 2.1 shows the level of congestion in London in the early 1900s, whilst Figure 2.2 is an aerial view of a modern-day motorway in the USA. A lesson learned during the 1980s and 1990s in the UK was that continual motorway building and road growth did not reduce congestion, as road space was rapidly filled upon completion.

Box 2.1 summarises some of the key dates that chart the early development of the motor car.

BOX 2.1 Key events in the industrialisation of the motor car

1859 Accidental discovery of oil (whilst searching for water in Pennsylvania, USA)

1860 Etienne Lenoir patents the spark-ignition engine

1876 August Otto produces first commercial four-stroke engine

1892 Rudolf Diesel patents the compression ignition engine

1908 First Ford 'Model T' sold in the USA

1913 Mass production of the Ford 'Model T' on the first modern production line

1924 MAN produces five-litre diesel engine for road vehicle use

1925 Automotive sector becomes the largest industry in the USA

1940 Over 200 cars per 1000 persons in the USA

1960 Global car population exceeds 100 million vehicles

1974 Clean Air Act passed in USA

1992 Auto-Oil programme leads to first European vehicle emission standards (Euro I)

1997 Toyota Prius becomes world's first commercially mass-produced petrol-hybrid car

2004 Global car population surpasses 600 million vehicles

2005 European Union introduces fourth round of vehicle emission standards for new passenger cars (EURO IV)

Figure 2.1 Congestion on a London road in 1919

Figure 2.2 Aerial view of a modern congested motorway – a Californian freeway, USA

With the additional development of small diesel power units, and the discovery of large reserves of crude oil in many regions of the world, the automotive industry expanded rapidly throughout the 20th century. During the last 50 years alone, the global vehicle population has increased by an order of magnitude to over 800 million vehicles (see Figure 2.3). A similar increase has occurred in the UK, where the number of registered road vehicles in use exceeds 30 million (Davis and Diegal, 2005). If the current rate of growth continues (at around 2–3% per annum), the global vehicle population could exceed 1 billion by the year 2012. There is, therefore, every likelihood that the motor vehicle will continue to significantly affect all aspects of life in the modern world.

Today, the transport and petroleum sectors have grown to a point where air and surface transport account for over 60% of global oil consumption and around a quarter of total energy consumption (IEA, 2005). The most common fuels for use by *road* transport are petrol and diesel, which are derived almost totally from crude oil. Both petrol and diesel require dedicated engine technology to convert the energy of these fuels into motive power. This chapter, therefore, begins by discussing the difference between petrol and diesel ICEs.

Figure 2.3 shows the growth of the world passenger car fleet as a function of time. Motorisation levels for future years, 2010 and 2015, are forecast using historical growth values of approximately 2% per year; these are shown in paler colours.

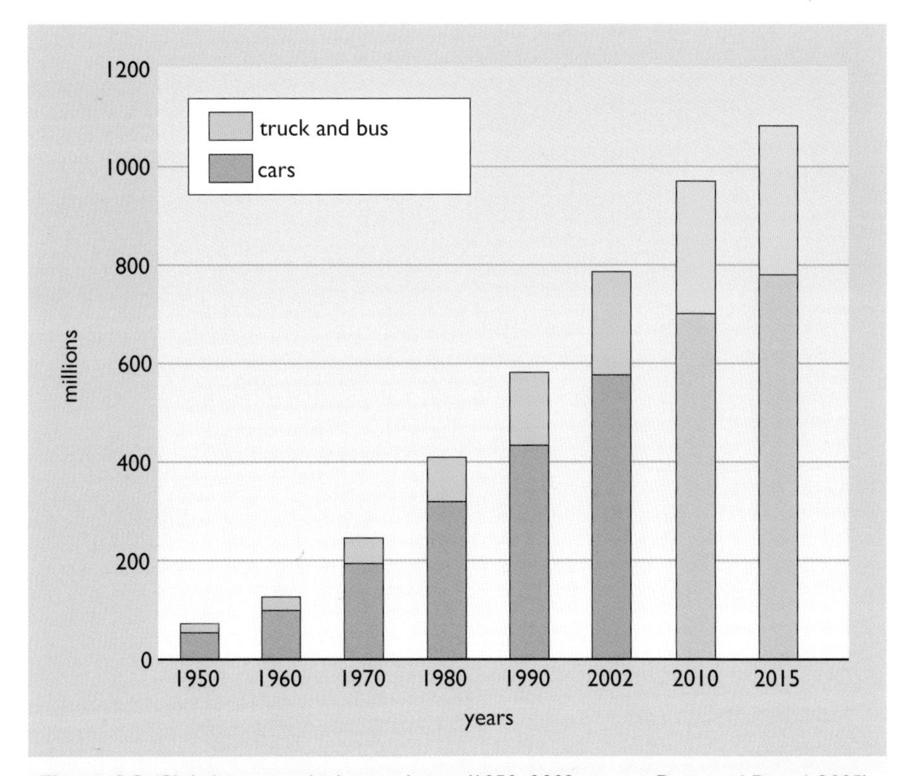

Figure 2.3 Global motor vehicle population (1950–2002, source: Davis and Diegal, 2005)

2.2 Petrol and diesel engines

The petrol-fuelled **spark-ignition** or 'Otto' engine (named after its inventor) is characterised by the use of a spark plug to initiate the combustion process. The engine utilises a four-stroke cycle, comprising the **induction**, **compression**, **power** and **exhaust** strokes. The four-stroke cycle is shown in Box 2.2.

BOX 2.2 Four-stroke and two-stroke engines

Modern petrol engines take two main forms: the four-stroke, using the Otto cycle, and the two-stroke cycle.

Both of these use a piston which is driven up and down inside a cylinder and connected to the drive section by a rotating crankshaft. At the top of a four-stroke engine there is a cylinder head containing a number of valves controlling the flow of gas in and out. The four 'strokes' are: **induction**, **compression**, **power** and **exhaust**. This is illustrated in Figure 2.4.

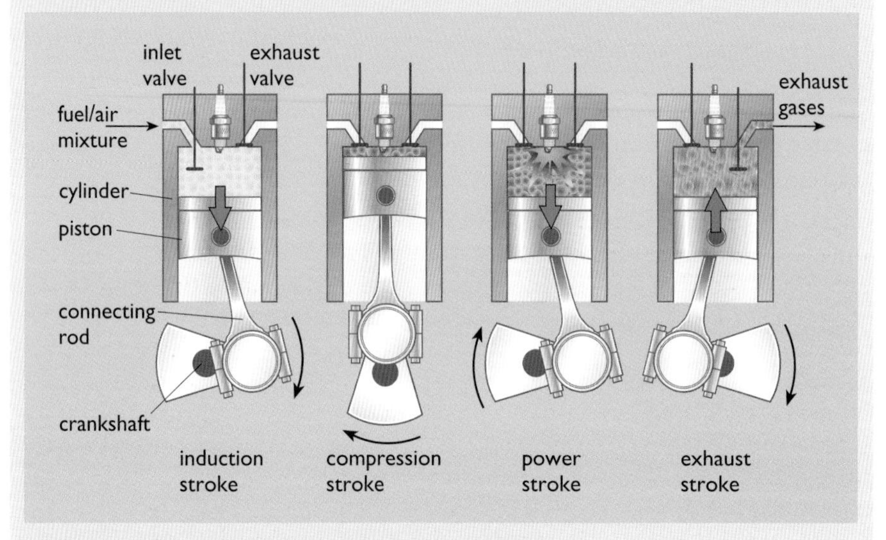

Figure 2.4 The four strokes of an Otto cycle engine

On the induction stroke a small amount of fuel and air is drawn into a cylinder through the open inlet valve, which then closes. On the next stroke this mixture is then compressed into a smaller volume. This reduction in volume is a rather critical factor called the **compression ratio**. In a modern car it is about 9:1, i.e. the fuel/air mixture is squeezed into one-ninth of its original volume, creating a highly inflammable mixture. This is then ignited using an electric spark on a sparking plug. The gases then burn very rapidly reaching a high temperature (750 °C or more) and expand, pushing down the piston on the power stroke. Finally, on the exhaust stroke, the burnt gases are pushed out into the exhaust system through the open exhaust valve. The whole cycle then repeats.

Starting with the induction stroke, a small amount of fuel and air are drawn into the cylinder cavity. Whereas older cars utilise a carburettor to mix the air and fuel to the correct ratio, modern vehicles tend to be equipped with fuel injectors, where the air intake is via a high-precision valve that sprays small amounts of petrol into the cylinder. This is usually done under the control of an electronic control unit (ECU). By using an on-board computer, the fuel injectors can vary the amount of petrol and air in order to achieve the lowest possible fuel consumption, or highest power output, depending on the engine load and accelerator position.

During the compression stroke, the petrol–air mixture is compressed into a small volume, usually to about a ninth of the original cylinder volume. In technical terms, the petrol engine is said to have a **compression ratio** of 9:1. (Typically, the air-to-fuel ratio is around 15:1 by mass.) The increase in pressure creates a highly explosive mixture, which is ignited by an electric spark (from the spark plug). The gases burn very quickly causing a rapid expansion and a release of chemical energy, which pushes the piston towards the connecting rod and crankshaft. This is the power or combustion stroke. Finally the burned gases, which ultimately make up part of the exhaust, are flushed out of the cylinder during the exhaust stroke via the exhaust gas port or valve. The cycle then starts over again with another induction stroke.

Higher compression ratios of 13:1 or more are possible using petrol, with careful engine design or by the use of fuels with a high octane rating, such as ethanol, methanol, natural gas or hydrogen. These can allow a higher combustion temperature and increased engine efficiency.

The diesel engine works using the same four-stroke cycle as the petrol engine, but with two major differences involving the air–fuel mixture and injection systems. In the diesel engine, only the air is compressed in the cylinder instead of an air–fuel mixture, and at the end of the compression stroke, the fuel is directly injected into the combustion chamber by a fuel injection pump. Typical compression ratios of 20:1 are used, which is sufficient to raise the air temperature to over 400 °C. Once the diesel fuel is injected into the cylinder, it immediately vaporises and spontaneously ignites. This combustion process produces a mixture of hot gases that then drive the piston. Diesel combustion is more explosive than petrol combustion. This leads to the characteristic diesel engine sound and explains why diesel engines are noisier and vibrate more than their petrol counterparts.

Figure 2.5 shows a modern petrol engine, clearly showing the pistons and camshaft which are visible in the cutaway section. The exhaust port is also shown on the left-hand side. This engine is an all-aluminium block with a displacement of 5.7 litres (346 in^3) and delivers a powerful 302 kW (405 hp) for a demanding US market. The engine is used across a range of models, mainly higher-end vehicles, and not surprisingly the next engines are being increased to 7.0 litres yielding 373 kW (505 hp) (GMC, 2003).

Modern diesel engines tend to use **direct injection** fuel delivery systems as they can be closely controlled by the use of computerised engine management systems. However, there are still many **indirect injection** diesel engines in the vehicle fleet. In these the fuel is injected into a pre-chamber before entering the cylinder. This allows for increased swirling (or mixing) that improves the combustion process. Older diesels also tend to be equipped with glow plugs, which heat the compressed air during a cold start by the use of an electrically-heated wire.

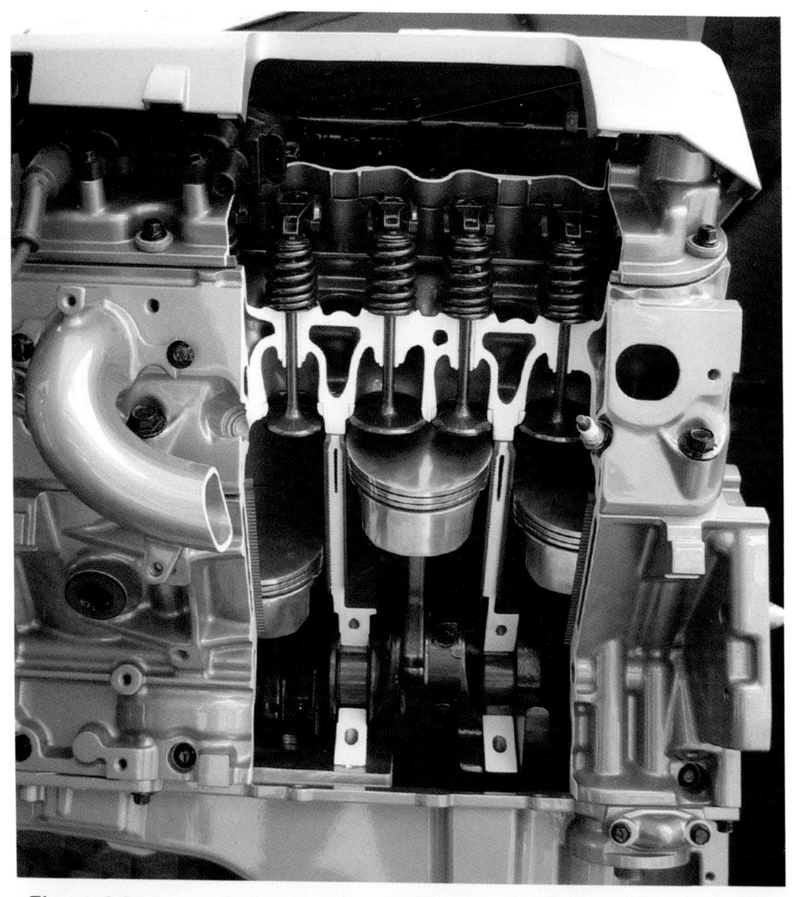

Figure 2.5 Cutaway of the Cadillac LS6 V8 injection engine

As the popularity of diesel engines has increased, several varieties of injection method have been developed, including **common-rail injectors** and **electronic unit injectors**. In the case of common rail, a single 'rail' or pipe is held at constant high pressure over the cylinders and a central control unit allows each injector to inject fuel electronically. Most systems use a pressure of about 1350–1500 bar to create two distinct fuel pulses: a pilot injection and the main (combustion) injection. The pilot injection helps seed the combustion process and can also be tuned to reduce engine noise. The next generation of engines will raise this pressure to a range of 1800–2200 bar along with multiple injections in order to lower emissions, engine noise and increase fuel efficiency without loss of overall engine power output. Electronic unit injectors are highly compact injectors that incorporate the fuel injection pump, the injector and the solenoid valve into a single unit. Further developments include the use of piezoelectric activators to improve injection control.

Further advances in engine fuel injection technology are expected over the next few years, due to increasing demands on the engine technology to be cleaner, quieter and more fuel-efficient. This may include the development of **homogenous charge compression ignition (HCCI)** engines, in which fuel and air are mixed before combustion thereby allowing a more uniform burn (Wells, 2006). A high compression ratio and very lean mixture (high ratio of air to fuel) increase energy efficiency and reduce emissions of nitrogen oxides (collectively called NO_x) – compared with conventional petrol and direct injection diesel engines. HCCI engines can be scaled to almost any size or application and can operate on a wide range of conventional and non-conventional fuels.

BOX 2.3 Commonly used transport units

Power and energy

kilowatt (kW), horse power (hp) – units of power produced by engines or electric motors

1 kilowatt = 1000 watts
1 hp = 746 watts

Kilowatt-hour (kWh) – unit of energy equal to one kilowatt operating for one hour

Joule – unit of energy equal to 1 watt-second
1 kWh = 3 600 000 J = 3.6 MJ (megajoules)

1000 MJ = 1 GJ (gigajoule)

kW kg^{-1} – power density; power produced by engine per unit weight
kWh kg^{-1}, MJ kg^{-1}, kWh $litre^{-1}$, MJ $litre^{-1}$ – energy density; energy contained in fuels or batteries

Pressure

Atmospheric pressure at sea level is approximately 1 bar, or 0.1 MPa in SI units.

Distances and speeds

1 mile = 1.61 kilometre (km)
1 km = 0.62 miles

50 mph approximately equals 80 kph

Fuel volumes and fuel economies

1 UK gallon = 4.55 litres
1 UK gallon = 1.20 US gallons
1 US gallon = 3.79 litres

mpg – miles per gallon (UK or US should be specified)

100 miles per (UK) gallon = 83 miles per (US) gallon = 2.83 L per 100 km

x L/100 km refers to the number of litres (*x*) of fuel required to travel 100 kms, for example 3 L/100 km is equivalent to 95 mpg (UK) or 79 mpg (US).

Emissions

g km^{-1} – grams of pollutant emitted per km travelled, usually used in EU emissions standards

g $mile^{-1}$ – grams of pollutant emitted per mile travelled, usually used in US emissions standards

Micron – millionth of a metre (also called micrometre)

PM_{10} – particulates up to 10 microns in size

$PM_{2.5}$ – particulates up to 2.5 microns in size

ppm – parts per million used in measurement of impurities, e.g. in fuel, but also for certain air pollutants for air quality measurements

ppb – parts per billion. Used in a similar way as ppm.

In general, the fuel efficiency of a diesel engine is higher than that of a petrol engine. This is primarily due to the fact that the combustion temperature (and pressure) within a diesel engine is higher than in a petrol power unit. This increases the engine's efficiency according to Carnot's equation for a perfect heat engine. This higher combustion temperature also leads to different exhaust emission profiles between vehicles with a petrol engine and those with a diesel engine (see Section 2.4). In addition, although diesel fuel has almost the same energy content per kilogram as petrol, it is more dense so it contains more energy per litre (see Table 2.1).

In a diesel engine about 32% of the heat energy is delivered to the crankshaft, whereas in a petrol car only about 24% becomes delivered work. As this kinetic energy is delivered to the wheel via the mechanical **drive-train**, energy is 'lost' owing to friction between the transmission components and to aerodynamic drag. As a result, only about 24% of diesel fuel's energy ends up being used for moving the car. In the case of petrol this is only 18%. Clearly, the actual values found vary enormously with the vehicle type and with the driving conditions (e.g. urban versus motorway driving). These figures could be considered relatively low, given the effort and cost associated with obtaining the fuel and manufacturing the vehicle in the first place. If we consider how much of the fuel's energy is actually used to move the payload, the situation is even worse. Taking into account the vehicle's mass, only around 1–2% of the fuel's energy is utilised to move the driver, passenger or freight.

Diesel's higher fuel economy (as compared to petrol) has been one of the reasons for the increasing demand (within Europe) for diesel cars. Other reasons include the facts that diesel engines provide increased low-end torque (more power at low engine revolutions) and higher peak engine power ratings. There is also a perception that diesel units are more durable than petrol power units. This **dieselisation** has seen an increase in the proportion of diesel cars in Europe from a base of 16% in 1985 to over 50% in 2006. This has important implications, not only on energy consumption and fuel distribution, but also on local air quality (See Section 2.4 on vehicle emissions).

Figure 2.6 shows the historic trend for increasing dieselisation of the passenger car fleet in both the UK and the EU-15 countries as a function of time (1995). Although the UK has a much lower diesel market sales penetration (approximately 36%) compared to, for example, France (approximately 70%), the market in the UK will probably continue to grow over the next few years, perhaps eventually achieving a market penetration of around 55–60% (Price WaterhouseCoopers, 2006).

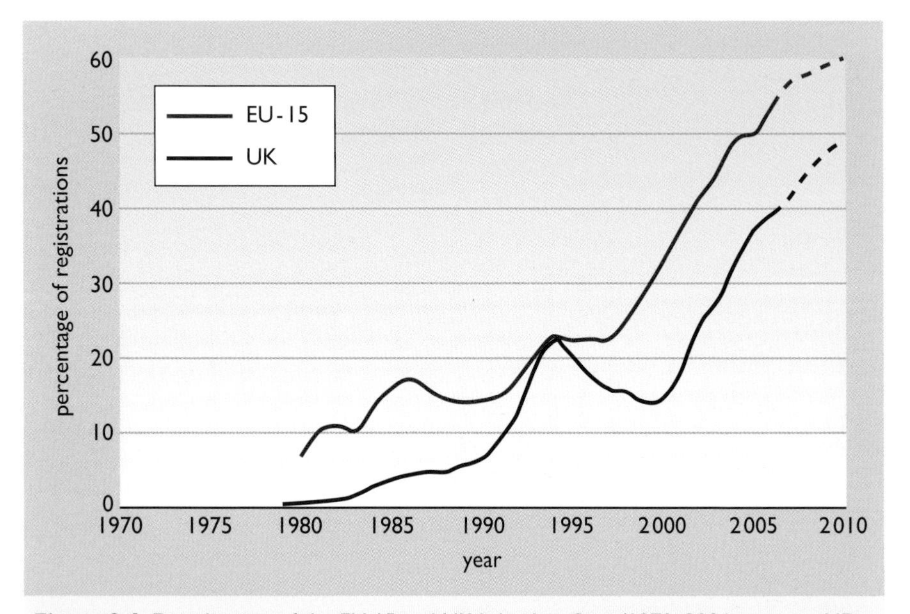

Figure 2.6 Dieselisation of the EU 15 and UK light-duty fleet (1970–2006, sources: AID, 2005; DfT, 2005; EC, 2005)

2.3 Petrol and diesel fuels

Petrol (known as gasoline or 'gas' in the USA) and diesel are mixtures of liquid hydrocarbons refined from crude petroleum. The production of these fuels involves the extraction of crude oil, separation from other fluids, transport to refineries, processing (fractional distillation), transport to regional storage locations and distribution to retail or fleet refuelling stations. Each fuel must be carefully blended, either to control petrol's volatility and anti-knock performance (**octane** number) or diesel's ignition quality (**cetane** number).

Diesel fuel's main hydrocarbon chain is $C_{14}H_{30}$, and petrol has a main component chain length of C_9H_{20}. More energy is required to 'crack' crude petroleum to produce shorter chains of hydrocarbons. This explains why diesel needs less energy to refine than petrol (only about half as much) and why petrol has a lower **viscosity**. Table 2.1 compares typical properties of the two fuels.

Table 2.1 Properties of petrol and diesel fuels

Fuel property (units)	Petrol	Diesel
Hydrocarbon chain length	C_4 to C_{12}	C_3 to C_{25}
Carbon content by mass (%)	85–88%	84–87%
Fuel density (kg litre^{-1})	0.75	0.83
Lower heating value (MJ kg^{-1})	43.2	43.1
Lower heating value (MJ litre^{-1})	32.2	35.9

Sources: Concawe, 2004; AFDC, 2003

Internationally, there has been a trend to introduce cleaner conventional fuels through the removal of lead, sulphur and other additives and impurities. For example, whereas lead was added as an octane rating improver in 1923, unleaded petrol was introduced in 1986 in the USA, with many countries following thereafter (Kitman, 2000). Indeed, owing to proven health risks, leaded fuels have been banned in the EU since 2000.

European fuel specifications have also led to reduced sulphur and polyaromatic content. These include ultra low sulphur diesel (ULSD) and ultra low sulphur petrol (ULSP). Since 2005, all petrol and diesel fuels sold in the EU have had to qualify as ULSD or ULSP, with a maximum sulphur content of 50 ppm. Previous specifications allowed up to 500 ppm sulphur content. The EU has also announced the mandatory introduction of 'sulphur-free' petrol and diesel (which in practice means a maximum of 10 ppm) by 2009.

Two of the first countries to introduce ultra low sulphur fuels were Sweden in 1991 and Finland in 1993. The Swedish and Finnish governments supported the introduction of these fuels through the use of differential fuel duties. In other words, the cleaner fuels were charged less tax than standard fuels as an incentive to consumers to buy the cleaner grades. The UK's fuel sulphur standards have also been more stringent than the EU standards – with the introduction of ULSP and ULSD being promoted through the levying of lower fuel duty (typically 1–2p per litre below other blends) to compensate for slightly higher fuel production costs. A further 0.5p per litre differential is intended to promote the switch to sulphur-free fuels, well ahead of the EU's 2009 deadline.

The main motivation for introducing reformulated fuels has been to reduce vehicle emissions. The reduction of sulphur in fuel significantly increases the longevity and efficiency of emission control systems (see Box 2.5) and reduces particulate emissions from diesel vehicles. However, removal of sulphur (and other impurities) is often associated with an increase in **production emissions** and processing costs. These production impacts must therefore be taken into account in assessing the merits of a reformulated fuel (see Box 2.6).

2.4 Petrol and diesel vehicle emissions

During the firing of the first diesel engine in 1893, the inventor himself noted that '...black, sooty clouds came from the exhaust pipe in all of these tests' (Monaghan, 1998). Rudolf Diesel would have to work for another four years before this problem was addressed and, even today, the issue of particulates from the use of the combustion engine remains.

Conventional road transport leads to environmental pollution as a result of physical and chemical processes which occur during vehicle and fuel manufacture, production, use, recycling and disposal. As a rule, the energy consumed during a vehicle's manufacture is relatively small (around 10%) in comparison to its energy use during its lifetime (Teufel et al., 1993; Mildenberger and Khare, 2000; Ecolane, 2006). Therefore, this section focuses on the emissions associated with vehicle operation, which includes the impacts of fuel production and use (see Box 2.6). (There are also environmental impacts associated with road construction, road maintenance and the development of the transport and fuel infrastructure required by a road-based transport system. However, these are not considered in this text.)

Within an internal combustion engine (in use), chemical processes take place between the hydrocarbons (HCs) of the fossil fuel, the fuel additives and the gases that naturally occur in the atmosphere (predominantly oxygen and nitrogen) (Boyle et al., 2003). The processes include complete and partial oxidation of the fuel, which produces carbon dioxide (CO_2), water (H_2O) and carbon monoxide (CO). Nitrogen from the air is also oxidised to nitrogen oxides (NO_x). Partially burned and unburned fuel are present in the exhaust gases and form a complex cocktail of volatile organic compounds (VOCs) together with small particles of matter ('particulates' or PMs). Tropospheric (low-level) ozone (O_3) is produced by the chemical action of sunlight on the VOCs, and subsequent reaction of the products with oxygen in the air. In those countries that still permit the use of 'leaded' petrol, lead (Pb) is also emitted with the exhaust gases. Box 2.4 provides a summary of the environmental and health effects of these emissions.

Petrol and diesel engines differ in their relative emissions performance, with petrol vehicles emitting fewer NO_x and particulate emissions, and diesel vehicles producing less CO_2 per kilometre. As NO_x production is predominantly associated with reaction temperature, the relatively high ignition temperatures attained during combustion can explain a diesel's higher NO_x emission. A diesel's lower CO_2 emission is due to the engine's higher efficiency as compared to petrol. Particulates up to 10 microns in size (termed PM_{10}) are also higher for diesels, although ongoing research suggests that petrol may produce more particulates in the $PM_{2.5}$ range.

BOX 2.4 Environmental and health effects of emissions associated with road transport

Carbon monoxide	During respiration, carbon monoxide combines with haemoglobin in the blood, which hinders the body's ability to take up oxygen. This can cause and aggravate respiratory and heart disease
Nitrogen oxides	Responsible for acid deposition via the formation of nitric acid. Dinitrogen oxide (N_2O; also known as nitrous oxide) contributes to global warming, and nitrogen dioxide (NO_2) is toxic to humans
Particulates	Responsible for respiratory problems and thought to be a carcinogen. According to the World Health Organization, there are no concentrations of airborne particulate matter (of size PM_{15} or less) that are not hazardous to human health
Volatile organic compounds	Benzene and 1,3-butadiene are both carcinogens and are easily inhaled owing to their volatile nature. Other chemicals in this category are responsible for the production of ground-level ozone, which is toxic in low concentrations. Also methane, released during the extraction of oil and during the combustion of petroleum products, is a powerful greenhouse gas
Carbon dioxide	The main environmental effect is as a greenhouse gas. Although there is uncertainty about exact numbers, the Tyndal Centre for Climate Change Research predicts a 1–2 °C rise in global temperature over the next 100 years (Lenton et al., 2006) due to increased CO_2
Tropospheric ozone	In the stratosphere, ozone absorbs ultraviolet light, therefore reducing the number of harmful rays reaching living organisms on the earth's surface. However, at ground level, ozone is toxic and responsible for aggravating respiratory problems in humans and reducing crop yields
Lead	Lead is known to affect the mental development of young children and is toxic in small quantities. Originally introduced into petrol to improve its octane rating.

Defra, 2002

Figure 2.7 compares petrol and diesel emissions from a typical passenger car with an engine size in the 1.5 to 2.0 litre range. These are measured in laboratory conditions, over a defined driving cycle, which represents a typical urban and extra-urban car journey, in a controlled laboratory. The values quoted are an average of all the small-vehicle data in the source and are not weighted to account for sales volumes. Note the relative levels of CO, NO_x, PM and CO_2. To some extent, the emission profiles of petrol and diesel illustrate the general tendency for different conventional technologies to 'trade off' emissions against each other. In this case local pollutants (NO_x, PM) are traded off against global ones (CO_2). This inability of the internal combustion engine to significantly reduce all emissions simultaneously *may* be an indicator that the technology is reaching the final stages of its development. However, only time will tell whether the ICE will be superseded by an alternative energy conversion device (see Sections 2.6 to 2.10).

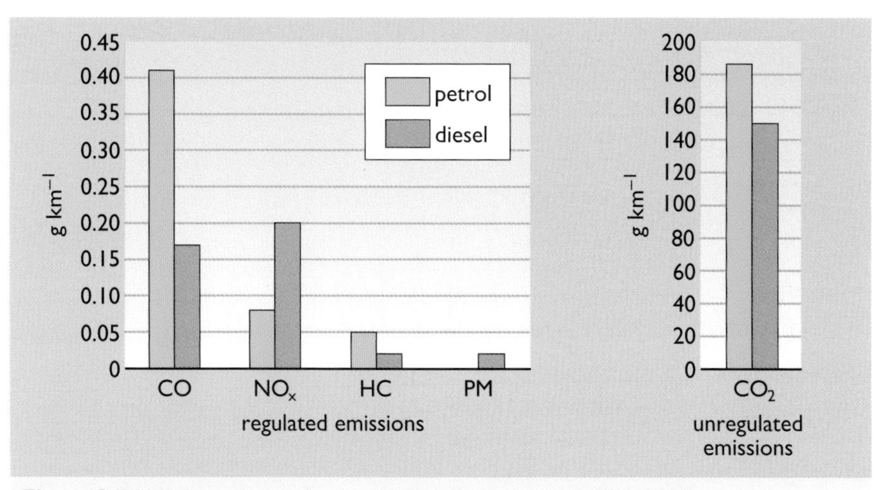

Figure 2.7 Vehicle emissions for a typical small car (source: VCA, 2006)

During the last 30 years, several technological advances have significantly reduced the emissions from ICE vehicles. One of the most important emission control technologies has been the introduction of the **three-way catalytic converter** (see Box 2.5 and Boyle et al., 2003). These were first used in the USA in the 1970s so that vehicles would conform to the Clean Air Act, one of the first regulations that limited pollution from mobile (and stationary) sources. As a *technical fix*, these catalyst systems have done much to improve air quality over the years in the USA, Japan and Europe.

BOX 2.5 **Catalytic converters**

Catalytic converters are an important type of 'end of pipe' technology that reduces emissions of CO, NO_x and unburned HCs from the exhaust of petrol engine vehicles (and are hence known as 'three-way' catalysts). Catalytic converters use a mixture of platinum, palladium and rhodium metals as their active components. In the presence of air, these catalysts promote chemical reactions that convert emissions to less harmful gases. The catalysts are applied to a high-surface-area support structure (within the exhaust pipe) through which the exhaust gases are made to flow. The units are protected in a steel or metal canister, located within the vehicle's exhaust pipe.

Most systems have to meet stringent durability requirements including working for 100 000 km or 5 years – whichever occurs first. Converters do have some inherent drawbacks. They are relatively ineffective before the 'light-off' temperature is reached, which means that they are inactive during short trips. Also, they tend to slightly increase fuel consumption (and hence CO_2 emissions). The precious metals in the converters can also be poisoned by certain fuel components such as lead and sulphur, which is why the use of catalysts has been dependent on the availability of lead-free and low sulphur fuels.

As in the USA and Japan, European legislation continues to be tightened for vehicle emissions (see Table 2.2) and has been highly successful in reducing some of the pollutants associated with road transport. In Europe, the Auto-Oil programme (a tripartite project involving the European Commission, oil and motor industries) has led to the introduction of mandatory limits for what are termed the **regulated emissions**. These are carbon monoxide, nitrogen oxides, hydrocarbons and particulate matter less than 10 microns in size (PM_{10}). In particular, key legislation (for passenger cars) was introduced in 1992 (known as Euro I), in 1997 (Euro II), in 2001 (Euro III) and in 2006 (Euro IV) (see Table 2.2). Similar European limits have been introduced for heavy-duty vehicles (specified in terms of grams per kWh of engine output).

Table 2.2 Past, current and future European emissions limits for passenger cars

Emissions Limits	**Petrol (g km^{-1})**				**Diesel (g km^{-1})**			
	CO	HC	NO_x	HC + NO_x	CO	NO_x	HC + NO_x	PM
Euro II (1997)	2.20			0.50	1.00		0.70	0.080
Euro III (2001)	2.30	0.200	0.15		0.64	0.50	0.56	0.050
Euro IV (2006)	1.00	0.100	0.08		0.50	0.25	0.30	0.025
Euro V (proposed 2009)	1.00	0.075	0.06		0.50	0.20	0.25	0.005

Source: DieselNet, 2006

It is interesting to note that, although transport is responsible for around a fifth of CO_2 emissions in the UK, there is no legislation, to date, that limits the amount of CO_2 produced per km for road vehicles. However, the European Commission's target is to reduce emissions of CO_2 from new cars sold in the EU to an average of 140 g km^{-1} by 2008 and 120 g km^{-1} by 2012. This would represent a cut of around 25% of the current average. To achieve this aim, the Commission has reached a formal (though voluntary) agreement with ACEA (the European car manufacturers' representative organisation) to implement the required technologies to reduce carbon emissions.

BOX 2.6 **Life cycle analyses of vehicles**

The **life cycle analysis** of road transport emissions is an established methodology that has been used by many researchers to compare the environmental impact of different road vehicle fuels and technologies (MacLean and Lave, 2003).

A full analysis of road transport emissions needs to account for both **vehicle emissions** generated during vehicle operation ('tank-to-wheel') and **production emissions** generated during fuel production, processing and distribution ('well-to-tank'). The total emissions due to fuel production and vehicle use are termed **fuel life cycle emissions** (also known as 'well-to-wheel').

Emissions (and other environmental impacts) are also associated with *vehicle manufacture*. Though they are not insignificant, they are not usually included in a comparison of fuels or vehicle technologies, unless it is thought that the difference in vehicle production methods is very different from conventional manufacturing processes. This can be the case where radically new technologies are being considered (e.g. change from ICE to electric vehicle technology). However, for conventional vehicles emissions associated with vehicle manufacture are not included in this chapter, as they are only a small proportion of the total life cycle emissions (Ecolane, 2006).

Although for petrol and diesel vehicles the fuel energy costs over a vehicle's life are far larger than the energy used in manufacturing it, this may not hold true for battery electric vehicles, where the energy costs of manufacturing, and recycling lead acid batteries can be significant. Therefore emissions generated in the disposal of a vehicle, including its component parts, must be considered.

Note that vehicle emissions are specified in grams per kilometre for light-duty vehicles and grams per kWh (engine output) for heavy-duty vehicles. Production emissions are specified in grams per unit of energy delivered (usually in gigajoules).

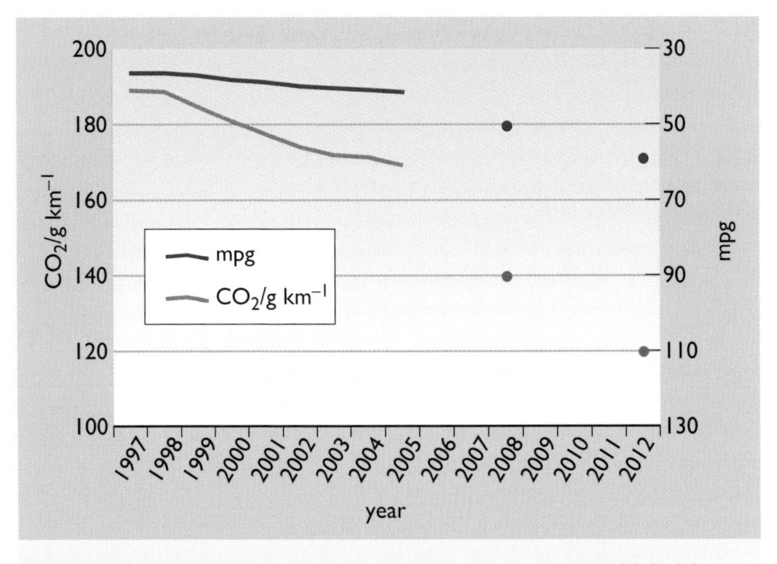

Figure 2.8 Vehicle fuel economies and CO_2 emissions in the UK (solid lines). Future target mpg values and corresponding CO_2 values assume a 50% dieselisation factor in the UK (single points) (sources: 1997–2006, SMMT, 2006; target values, ACEA, 2003)

For petrol and diesel vehicles, carbon emissions are closely correlated with fuel use. Therefore, the trends in CO_2 emissions are similar to those for fuel economy. Given the facts that engine designs are becoming more efficient, and that cars utilise more lightweight materials and are becoming more aerodynamic, one might think that the overall fuel economy is improving and CO_2 emissions, on average, are decreasing. However, this is not the entire story. Engine power is increasing, driven by consumer demand for more powerful cars, and for more extra features within the vehicle. Equipment such as air conditioning, heated seats, electric windows, auto-defrosting, and on-board navigation all require energy to operate, and can result in increased vehicle weight. As of 2005, the average new (petrol-powered, two-wheel drive) car fuel consumption was 6.8 litres per 100 km (or 41.4 mpg) which is equivalent to around 170 g of CO_2 per km. This represents an 11% improvement over 1997 values, a change due predominantly to the ACEA agreement. However, as Figure 2.8 shows, there is a long way to go to reach the 2008 and 2012 targets and some uncertainty about whether these targets can be achieved (LowCVP, 2006). Figure 2.9 depicts fuel consumption and targets, and shows the relationship between L/100 km and mpg. With respect to the 3L/100 km target, some vehicle manufactures have been mapping the possibilities of 1L/100 km (which is more than 280 mpg).

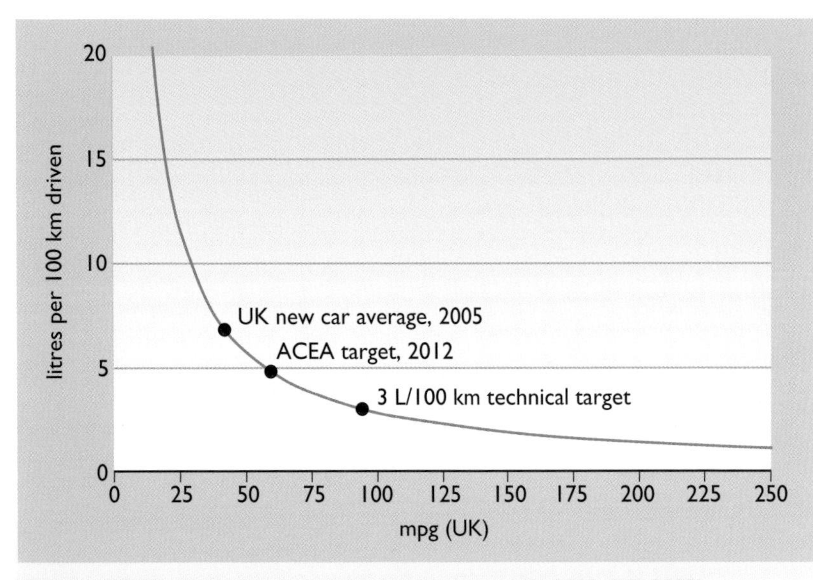

Figure 2.9 Fuel consumption and targets (source: AID, 2005; DfT, 2005)

2.5 Cleaner conventional vehicle technologies

Vehicle emission legislation has been one of the strongest factors forcing car manufacturers and their suppliers to develop less-polluting engines. Technology improvements to date include more efficient engine designs, new tail-pipe emission control and electronic management systems, and improved sensing devices to monitor the state of the engine and exhaust.

For some time, some manufacturers have been developing the next generation of petrol power units, which include gasoline direct injection (GDI) engines. First developed by Mitsubishi, this technology offers up to 20% fewer carbon emissions and a similar improvement in fuel efficiency. A GDI engine works like a normal petrol power unit, except that the petrol is sprayed directly into the combustion cylinder (there is no pre-mixing stage as in a typical petrol engine). This results in a cleaner burn and an increase in power. The main obstacle to this technology is the sulphur content of petrol, which, even at the levels found in ULSP (50 ppm), hinders the catalysts that are required to control the NO_x emissions.

In addition to the three-way catalytic converter, much work has been conducted to develop new exhaust emission control systems for petrol and diesel engines. One such device is a **particulate filter**, fitted to an increasing number of new diesel vehicles. This is a complex system containing a filter to trap the soot, an active fuelling strategy that helps burn the trapped particles and a control system to monitor the soot level initiating combustion of the particulates when required. For heavy-duty engines, emission control devices include oxidation ('one-way') catalysts, exhaust gas recirculation (EGR) systems, selective catalytic reduction (SCR) systems and continuously regenerating traps (CRTs). These devices are increasingly being fitted as standard and are proven to reduce particulates by up to 90%. These technologies are required for many diesels to comply with Euro IV standards and for Euro V, which is due in 2009.

Table 2.3 shows some of the technology improvements that are likely to continue to be introduced for diesel and petrol ICE vehicles in the next decade. Although many of these advances will be effective at improving fuel use and reducing vehicle emissions, one technological breakthrough may prove to be even more productive (and cost-effective). This approach addresses one of the intrinsic incompatibilities of conventional engine vehicle use; namely that, for a conventional ICE vehicle, the maximum efficiency of the engine is achieved under certain conditions, usually at 2000–3000 engine revolutions per minute (rpm) when running at approximately 90% of full power at that speed (Stone, 1999). These 'perfect' engine conditions are rarely achieved during urban driving, which includes low speeds and stop–start traffic. As a result, the average engine efficiency falls far short of its design optimum (as do the fuel economy and vehicle emissions). To address this issue, some manufacturers are developing a new engine configuration which increases the time an ICE engine can operate close to its point of maximum efficiency; the **hybrid electric vehicle**.

Table 2.3 Future likely improvements in engine and emission control systems

Technology type	Diesel	Petrol
Increasing exhaust gas recirculation (EGR)	✓	✓
Higher injection pressures	✓	
Particulate trapping	✓	
Variable pressure turbocharging		✓
Cylinder deactivation		✓
Heated catalysts		✓
Engine downsizing (with electronically assisted turbocharger)	✓	✓
Use of lightweight components (aluminium, plastics)	✓	✓
Automated manual transmissions	✓	✓
Integrated starter/alternator units	✓	✓

Sources: adapted from Fraidl et al., 2000; Pearson, 2001

The hybrid electric vehicle

When an ICE vehicle is combined with a battery electric traction system, the result is called a hybrid electric vehicle (HEV). In a very real sense, a hybrid is part conventional ICE technology and part electric vehicle (see Section 2.9). The principle underlying all hybrid vehicles is the use of an energy buffer (usually a secondary cell, also known as a rechargeable battery) which enables the main power unit to be operated at close to maximum efficiency. When the engine loading is low, the excess output is stored as chemical energy (within the battery) for later use. When the loading is high, the main engine and the battery work together to deliver the required power. The use of an on-board battery also enables the use of **regenerative braking**, which recovers part of the energy usually 'lost' during braking, so reducing overall energy use. In this way, HEVs provide significantly improved fuel economy and reduced emissions.

Hybrid vehicles can be categorised as **parallel**, **series** or **split** hybrids (see Figure 2.10). Each of these definitions refers to the configuration of the main and peak power units. Briefly, the three different systems can be summarised simply as:

- **Parallel** – the engine and electric motor are both connected to the transmission, so that either the engine or the motor can provide power to the car's wheels
- **Series** – the engine does not directly provide power to the car's wheels; instead the engine drives a generator, which can power the electric motors that run the wheels or charge the batteries
- **Split** – the engine drives one axle whilst the electric motor drives the other. There is no connection between the mechanical and electric drive-trains. The split system is also referred to as the series parallel system.

A technical breakthrough in hybrid vehicles occurred during the 1990s, when several motor manufacturers developed HEVs to (or almost to) production stage. The first commercially available **petrol–electric hybrid**, the Toyota Prius, was launched in Japan in 1997 and in Europe in 2000 (Toyota, 2006). Hybrid passenger cars have also been launched by Honda (the Insight and Civic), Lexus (the RX400h and GS450h) and are being developed by other manufacturers including Nissan, Audi, Renault, Peugeot and Volkswagen. Diesel hybrid bus and truck vehicle projects are also under way in many European cities.

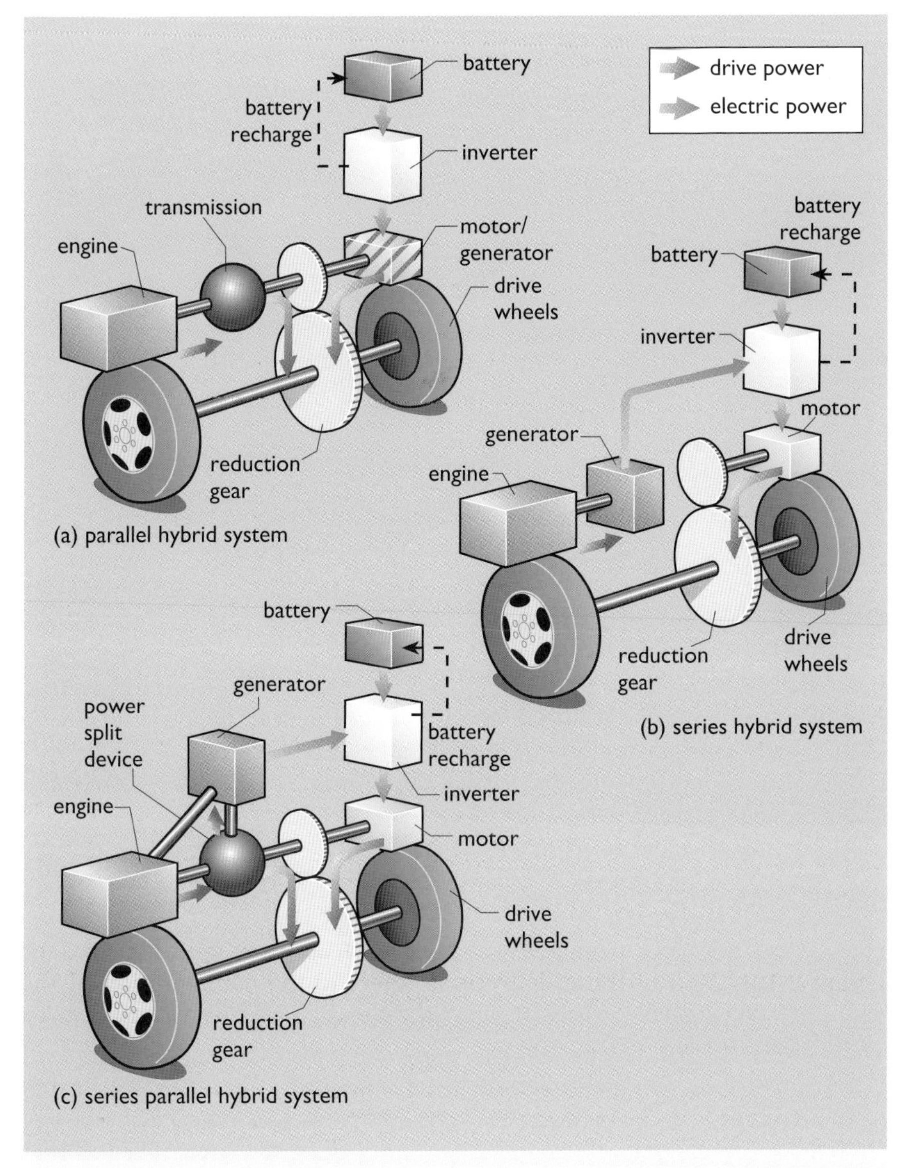

Figure 2.10 Types of hybrid vehicle

The Prius's hybrid system incorporates a 57 kW, 1.5-litre petrol engine rated at 5000 rpm and a 50 kW electric motor, yielding an overall maximum power of 82 kW. The Prius can be categorised as either a series or a parallel hybrid, owing to the unique nature of an electronically controlled 'power splitter'. This device allows the hybrid system to direct power from the conventional engine to either the wheels or to the generator. The generator in turn can drive a motor (to power the wheels) or to charge the battery. The battery is used to drive the motor when the engine is off or needs extra power.

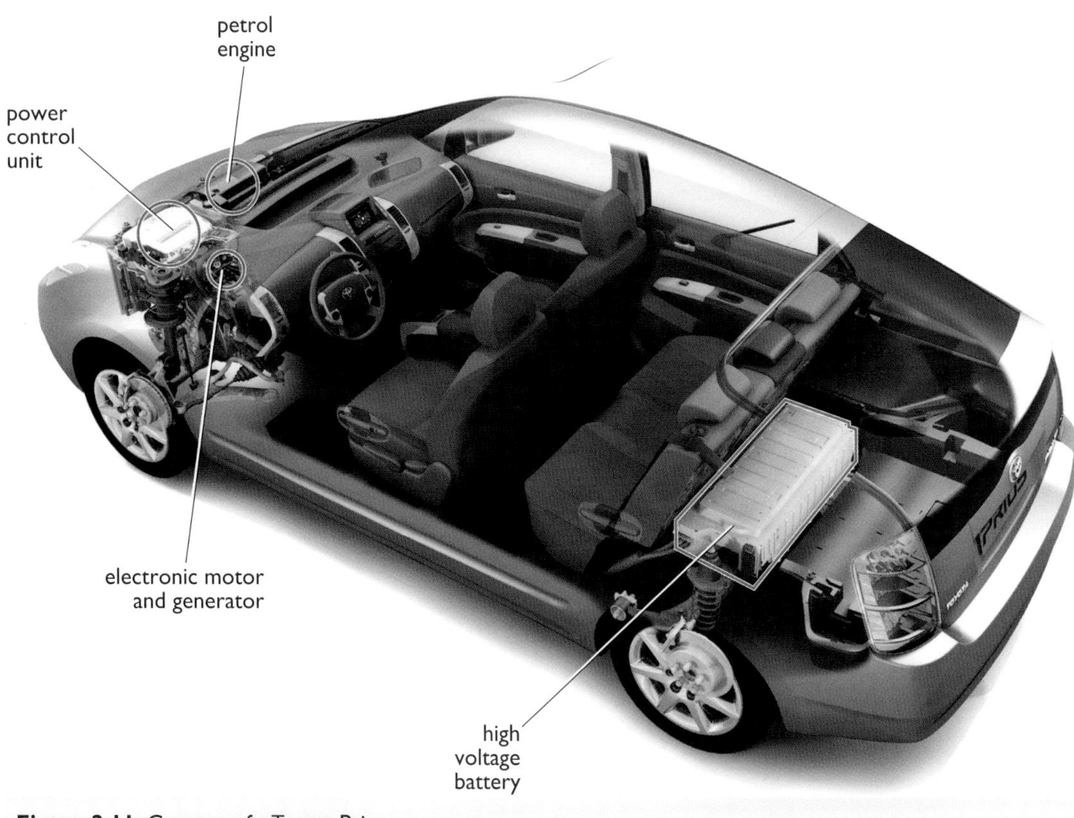

Figure 2.11 Cutaway of a Toyota Prius

By 2005, over a quarter of a million Priuses had been sold confirming it as the worlds's first commercially successful hybrid car, significantly outselling other hybrid models and other types of battery electric vehicle. Given this initial, though modest, success, many of its innovative features are likely to become standard in other hybrid vehicles over time.

Under test, the Prius shows that all regulated emissions are significantly reduced as compared to an equivalent petrol ICE vehicle. In 2000, the car already complied with the Euro IV standard (six years ahead of EU legislation). Research in the USA also shows that hybrids are more fuel-efficient than a conventional diesel vehicle (and emit less CO_2) by 20–30% (Cuddy and Wipke, 1997). This is confirmed by the Prius, which achieves a fuel economy of 4.2 litres/100 km (approximately 65 mpg) on a standard European drive cycle, dependent on the specification. This represents a fuel economy improvement of around 25% as compared to a conventional petrol car. Given that its CO_2 emissions are around 104 g CO_2km^{-1}, it seems that hybrids may provide the auto industry with a technology that can deliver the EU/ACEA targets for vehicle carbon emissions (see Section 2.4).

As with most new technologies, the benefits of hybrids come at a price; they are typically 15–20% more expensive than a conventional petrol equivalent. However, the Prius does reduce fuel costs and that offsets its higher capital cost, as the reduction of fuel use by 25% translates directly into a fuel cost saving of the same amount. In certain countries, government grants are available to offset the higher cost of hybrids. The grants range from 4–16% of the initial purchase cost, depending on country and final retail price.

Hybrid electric vehicles are still at an early stage of development and a dominant system design (if one exists) has yet to emerge. Thus, equal numbers of series and parallel systems are under development. Generally, there seems to be a consensus that parallel hybrids will dominate the market initially, followed by series designs. It is difficult to predict the future success of hybrid technology until the first commercially available vehicles have been used in real driving conditions over an extended period of time. However, early experience has been very encouraging and hybrids possess great potential to become the standard automotive technology during the coming decade.

One intriguing recent development is the after-market conversion of a Toyota Prius. Known as a 'plug-in hybrid' (see Box 2.7), the vehicle is given an additional battery pack that can be recharged from the mains supply like a battery electric vehicle, or it can be topped up by the engine along with the existing battery. The additional battery is a lithium–iron–phosphate-based system which allows a longer electric-only driving range and, according to the conversion company, increases fuel economy to 130 mpg (EAST, 2005). Hybrids have advantages and disadvantages; some of these are included in Box 2.7 (those of the other technologies discussed are given in boxes within the relevant sections).

BOX 2.7 The 'plug-in' hybrid and the advantages and disadvantages of hybrids

The plug-in hybrid

What can be done to show the British that the electric vehicle is a viable, economic and non-polluting alternative? One venture that aims to switch the public on to the efficacy of such vehicles is a company called 'greentomatocars.com'. This London minicab company's fleet is comprised of petrol–electric hybrids: 'We want people to see that electric cars are green and practical,' says co-founder Tom Pakenham. 'Our fares are the same as other companies', but our vehicles emit less than half the carbon dioxide of a traditional black cab.'

Pakenham sees his venture as a way of familiarising people with green-car technology. 'Some clients worry that if they book an electric car, it might run out of power on the way to the airport. So our drivers are all briefed to explain – when asked – how the hybrid uses the electric battery in slow-moving traffic, but switches to petrol on faster roads, recharging the battery at the same time.'

Of course hybrids burn petrol and emit CO_2, but at lower levels than equivalent vehicles. The Toyota Prius ... produces 104 grams per kilometre, less polluting than a small car such as a MINI (129 g/km) or the famous Smart car (113 g/km), and one-third of the CO_2 of the average 4x4. On the other hand, the hybrid Lexus belonging to the Conservative Party leader ... produces 184 g/km, while the [Chancellor's] government-issue non-hybrid Vauxhall Omega churns out 276 g/km.

Given that hybrids are greener, rather than green, vehicles, Pakenham wanted to run an all-electric fleet. Unfortunately that was not possible. 'There is no mid-size consumer electric car on the market,' he explains. 'I could have had them custom-built in the US, but at £50,000 each, we wouldn't have been able to offer the same fares as other taxi firms.'

So greentomatocars.com plans to make its fleet less polluting with upgrades from [a company called] Amber Jack. This involves the exchange of the Prius's nickel metal hydride batteries for lighter and more efficient lithium ion ones (the type used in mobile phones and laptops). They also fit a charger, so that the batteries can be connected to the domestic mains, and topped up overnight. Once converted, at driving speeds of less than 31 mph, the Prius will run for 70 miles without using petrol. The petrol engine only comes into play if you exceed 31 mph (a default set by the original manufacturer) or go beyond the 70-mile battery range without recharging.

The average town commuter may never burn any petrol with an 'adapted' hybrid, effectively turning it into an all-electric vehicle. But to gain the full benefits from the conversion, the vehicle will need access to a domestic 13-amp power point for the seven-hour recharge (cost of electricity: 22p). And the upgrade will void the vehicle guarantee.

The two-hour upgrade will be offered for the Prius from September [2006], and there are plans to make it available for other hybrids sold in the UK. Unfortunately, only motorists whose wallets can stretch to a £17,000 Prius and the £9,000 modification will be able to drive them. But once they have bought the upgrade, for every 10 000 miles driven [using mains electricity] they would save around £1000 at current fuel prices. And [if powered solely by 'green tariff' renewable electricity] they would be producing virtually no carbon-dioxide emissions.

Zakian, 2006

Advantages and disadvantages of hybrid vehicles

Advantages:

- **Lower overall fuel consumption** – for certain driving conditions and cycles
- **Lower life cycle impacts** – (i.e. greenhouse gases) due to lower miles per gallon
- **Electric motor output** – (specific power density kW kg^{-1}) is higher than for petrol engine alone, thus performance can be improved
- **Higher levels of torque and faster response** – greater acceleration than conventional cars, according to some manufacturers.

Disadvantages:

- **Reduced internal space** – hybrids need more room on-board for complex sub-units, which increase overall vehicle mass
- **Reliability and maintenance** – some consumers are uncertain about these issues for hybrid power trains
- **Capital cost** – hybrid production and purchasing costs are higher than conventional cars, although predictions say this might reduce with increasing scale
- **Battery disposal and resources** – larger additional hybrid battery requirements increase issues about resource depletion and disposal.

Adapted from Costlow, 2006

2.6 Alternative vehicle fuels and engines

As we have seen in previous sections, conventional vehicle systems are based on petrol and diesel fuels and on the internal combustion engine. During the past century, these fuels and technologies have become highly developed and are supported by global industries that have made vehicles affordable to most people in the modern world. There have also been improvements to engine efficiency during that time, which has allowed the addition of on-board devices that improve driver and passenger comfort and safety. From an environmental perspective, the regulated emissions have been significantly reduced.

However, limits to the development of the conventional ICE vehicle have become apparent. Despite continuing efficiency improvements driven by the ACEA agreement, car fuel economy has not improved significantly over the past few decades (see Figures 2.8 and 2.9) and average engine efficiencies remain at around 15–20%. Worse, as already mentioned, owing to the relatively high mass of vehicles, only a few per cent of petrol's or diesel's energy is utilised to actually move the driver or payload. The motor car is also totally dependent on the supply of crude oil, which makes the use of motor transport a highly political issue. Lastly, the transport sector is a significant contributor to total greenhouse gas emissions (around 20% in the UK), and even though regulated emissions are reducing **per vehicle kilometre**, NO_x and particulates remain a problem in many cities owing to the increasing number of vehicles on the road and miles travelled.

For these reasons, governments, together with the fuel and automotive industries, have been attempting to develop **'alternative' vehicle fuels**, which could reduce dependence on oil and/or lessen road transport's environmental impact. This process has started with the introduction of ultra low sulphur diesel and petrol. But truly alternative fuels may provide further benefits in the longer term, including those related to air quality and climate change. The advantage of many alternative fuels is that they can be used in 'conventional' ICE vehicles. Their use also improves the use of advanced after-treatment systems that can further reduce vehicle emissions.

Cleaner fuels that provide tangible emissions benefits include **natural gas, liquefied petroleum gas, biofuels** and **hydrogen.** These are already being used for transport applications and have been shown to reduce vehicle and **life cycle emissions** on a per-kilometre basis (see Box 2.6). In principle, they can be used in most ICE vehicles with relatively minor modification. Indeed, BMW has developed a series of hydrogen ICE prototype cars, the latest of which includes the hydrogen-powered MINI (Boyle et al., 2003).

A second (and more radical) strategy is to develop **'alternative' vehicle technologies.** These involve the use of totally new energy conversion systems that partially or completely replace the internal combustion engine. In fact, this process has already begun with the introduction of the **hybrid vehicle** (see last section), which, as the name suggests, combines the advantages of an ICE with that of an electric drive-train. Indeed, most of the alternative vehicle technologies under development employ electric (as opposed to mechanical) drive-trains. These include **battery electric**

vehicles, which are particularly suited for urban and short-range use, and the **fuel cell electric vehicle,** which is (usually) fuelled by hydrogen and which is considered by many in the motor industry to offer great potential as a road vehicle technology. Fuel cells have been used for space exploration since the 1960s (Boyle et al., 2003). Indeed, the more general **hydrogen economy** has been considered by some analysts to be 'inevitable', providing a means of long-term storage for renewable energy (Serfas *et al.*, 1991). Though hydrogen can be used within existing ICEs, **hydrogen fuel cell vehicles** would radically change patterns of transport energy use and environmental impact.

Table 2.4 shows the alternative vehicle and fuel technologies that will be discussed in the following sections. The options considered by no means form an exhaustive list, but they do represent the alternatives considered by most analysts to have the potential to be commercially viable within Europe by 2020.

Table 2.4 Alternative vehicle and fuel technologies

Alternative vehicle technologies (typically using electrical drive-trains)
Hybrid electric vehicles (HEV)
Battery electric vehicles (BEV)
Fuel-cell vehicles (FCV)
Alternative fuels (typically using mechanical drive-trains)
Compressed natural gas (CNG)
Liquefied petroleum gas (LPG)
Biofuels (bio-ethanol and bio-diesel)
Hydrogen

2.7 Compressed natural gas and liquefied petroleum gas

Compressed natural gas (CNG) and liquefied petroleum gas (LPG) are mixtures of low-boiling-temperature hydrocarbons. The main constituent of natural gas is methane (CH_4) with smaller amounts of propane (C_3H_8) and other hydrocarbon gases. LPG is a mixture of propane (over 90% in UK) and butane (C_4H_{10}). Being relatively simple chemical compounds that mix easily with air, these gases enable a more complete combustion than do conventional liquid fuels, which can lead to a reduction in vehicle emissions. The gases also have high **octane ratings** that enable a high compression ratio to be used, so improving engine efficiency.

Natural gas vehicles (NGVs) were first introduced in Italy just before the Second World War, for use in light-duty commuter cars, and were supported through the use of government subsidies. The Argentinian government were also early promoters of the fuel, partly in response to severe air pollution problems in Buenos Aires, and partly to conserve their own supplies of oil for export to earn foreign currency (IEA, 1999). Currently,

Table 2.5 Properties of alternative fuels

Fuel property (units)	CNG	LPG	Methanol	Hydrogen
Chemical formula	CH_4	C_3H_8	CH_3OH	H_2
Carbon content by mass	75%	82%	37.5%	0%
Typical storage pressure (bar)	200	8	1	200
Fuel density (kg litre^{-1})	n/a	0.51	0.80	n/a
Lower heating value (MJ kg^{-1})	47.6	46.4	19.9	120
Lower heating value (MJ litre^{-1})	n/a	23.6	15.7	n/a

Note that fuel density and volumetric heating value cannot be specified for gaseous fuels. Sources: DTI, 2000; AFDC, 2003

5 million NGVs are in use worldwide supported by a network of 1500 fuelling stations (ENGVA, 2006) with Argentina, Brazil, Pakistan, Italy and India operating the largest fleets. Europe has around 470 000 NGVs (400 000 of which are in Italy) serviced by over 2000 filling stations. Around five hundred NGVs are located in the UK, mainly operated by fleet operators in the private and public sectors (NGVA, 2006).

In the Netherlands, LPG is already considered a 'conventional' motor fuel, with most Dutch motorway filling stations supplying the fuel and around 6% of the light-duty vehicles using the fuel. Several major Dutch cities have public transport fleets operating on LPG and it is common for transport companies to buy buses for conversion to gas. Worldwide, there are currently over 10 million LPG vehicles, with 3 million in Europe alone; the largest fleets are in Italy (more than 1 million), Australia, North America and the Netherlands (LPGA, 2006; AEGPL, 2006). In the UK, there are over 120 000 LPG vehicles on public roads, the majority being light-duty vehicles that have been converted to run on LPG fuel.

Vehicle technology

Most light-duty vehicles that operate on 'road gas' fuels are **bi-fuel** conversions. These utilise a traditional spark-ignition petrol engine that can also run on LPG or natural gas. Whereas older conversions often had poor performance (the engine being optimised for petrol operation), more recent conversions incorporate fuel injection systems that have greatly improved engine response for both fuels. However, some drivers of bi-fuel vehicles continue to report some power loss when using gas. For this and other reasons, dedicated gas engines maximise the benefits that are offered by LPG and natural gas and can provide vehicle performance similar to conventional fuels. In many cases, improvements in engine performance are found for heavy-duty vehicle conversions to gas, including higher torque at low rpm and an extended engine life due to the cleaner fuel and reduced engine stress.

Compressed natural gas is normally stored on-board a vehicle in a pressurised tank at around 200 bar. Cars are typically fitted with a single cylinder that contains 16 kg of gas, equivalent to the energy of 23 litres of petrol. For steel cylinders, which are most common, the combined tank–fuel weight is about four times heavier than for petrol/diesel. This increases fuel consumption and reduces the payload that can be carried. Therefore dedicated NGVs tend to be heavy-duty vehicles where the extra weight and volume of the gas tanks is less of an issue. LPG can be liquefied more easily than natural gas and is stored as a liquid under moderate pressure (at 4–12 bar). As LPG storage tanks pose less of a space problem than do CNG cylinders, LPG has become very popular within the light-duty sector in the UK. Uptake has also been promoted through the low cost of conversion, the ease of refuelling and the increasing availability of the fuel.

Figure 2.12 Refuelling a postal van in the UK with compressed natural gas, from a dedicated commercial vehicle at the Post Office's own site

Emissions benefits are offered by the use of road-gas vehicles (see Box 2.8), but their use is associated with increased capital costs. For example, the additional costs for a CNG storage cylinder (for a heavy-duty vehicle) can be as high as £10 000. Even for cars, CNG adds 10–15% to the cost of a vehicle. Conversion to LPG is less expensive for light-duty vehicles, at around £800–1500 for cars and vans, and around £15 000–25 000 for bus conversions.

Fuel supply and infrastructure

Natural gas refuelling systems can either be 'fast-fill', using gas at 250 bar to refuel a vehicle within minutes, or 'slow-fill', which uses a compressor to 'trickle charge' a vehicle over several hours. LPG is dispensed as a liquid under moderate pressure in much the same time it takes to refuel a petrol or diesel vehicle. In the UK, while there are only around 30 CNG filling points, there are over 1200 LPG stations. This explains in part why LPG has become the more popular gaseous fuel for light-duty use. However, although public refuelling facilities can service more vehicles, depot-based refilling sites are playing an important role in the development of NG, LPG (and other) cleaner fuels. This is because fleets using alternative fuels can be more easily managed using centralised refuelling and support facilities.

Figure 2.13 City bus running on LPG with fuel tanks clearly visible on the roof of the bus

The high capital cost of NG refuelling systems also acts as a barrier to the uptake of road gases. For example, in Southampton £250 000 was required for a system to fast-fill a fleet of 16 buses. Infrastructure costs are less of a problem for LPG as the fuelling units operate at lower pressure than for natural gas, which again explains why the uptake for LPG has been initially greater than for CNG. For LPG, the increased cost of gas vehicles and fuel infrastructure is partially offset by the relatively low price of gaseous fuels, which have benefited from advantageous fuel duties set by national government (the UK fuel duty on gaseous road fuels was cut from 21p/kg to 9p/kg over the period 1998–2005.

Environmental impact

For light-duty vehicles, with the exception of hydrocarbons the regulated emissions are significantly reduced for gas-powered vehicles. Compared to petrol car emissions, NO_x is reduced by at least a third and particulates are virtually eliminated. Hydrocarbons are reduced for LPG vehicles, whereas these emissions can be increased for some NGVs owing to the presence of non-combusted methane in the exhaust gases (which has significant implications for greenhouse gas emissions).

For dedicated heavy-duty vehicles, as compared to diesel, the reductions are around two-thirds for both NO_x and particulates. Emissions of hydrocarbons are reduced by well over 50% for LPG, though are significantly higher for heavy-duty NGVs. However, over 80% of these HC emissions are composed of methane that can be almost eliminated from exhaust gases by the use of dedicated catalyst systems.

Energy use per km is slightly increased for gas operation as compared to conventional fuels. However, owing to the NG's and LPG's low-carbon content (see Table 2.5), vehicle CO_2 emissions (per km) are reduced. For light-duty vehicles, tests provide evidence of a 10–20% reduction of life cycle CO_2 emissions as compared to petrol operation. For heavy-duty vehicles, life cycle CO_2 emissions are comparable to diesel operation.

In assessing the full impact on global emissions, it should be remembered that methane is an important greenhouse gas. Therefore, for NGVs, the methane emissions from the vehicle, refining and distribution processes must be accounted for in the calculation of the effect on global warming. The result is that, for heavy-duty NGVs, total life cycle greenhouse gas emissions are comparable or *slightly increased* when compared to diesel operation. This situation will improve as more dedicated gas engines are brought on to the market with optimised methane catalysts.

BOX 2.8 Advantages and disadvantages of road fuel gases

Advantages:

- **Reduced emissions** – reduced NO_x, particulates and CO_2 (for cars)
- **Reduced fuel costs** – up to 30% lower fuel cost per km using LPG
- **Reduced low noise levels and engine vibration** – noise reduction from 68 dB to 60 dB for heavy-goods vehicles (60 dB is equivalent to a typical car)
- **High fuel availability** – LPG available at over 1200 refuelling stations.

Disadvantages:

- **Higher capital costs** – 10–15% higher vehicle costs for light-duty conversions and up to £25 000 additional costs for heavy-duty dedicated gas vehicles
- **Poor NG refuelling infrastructure** – although NG is available through the national grid, very few filling stations have been installed owing to high equipment costs
- **Reduced vehicle payload** – mass and volume of gas tank can reduce payload capacity of heavy-duty NGVs by up to 1 tonne
- **Vehicle restrictions** – some restrictions in use of LPG in confined spaces (tunnels, car parks) within Europe.

BOX 2.9 LPG vehicles: Oxfordshire Mental Healthcare NHS Trust

Oxfordshire Mental Healthcare NHS Trust operates health services from various sites throughout the county. Small passenger vehicles meet most of their transport requirements, as the Trust operates a regular minibus service as a non-stop shuttle between its sites. The bus service ferries members of staff to and from work and delivers cost and environmental benefits in reducing staff reliance on private car use.

This Transit minibus service travels around 44 000 miles a year using two drivers who provide the service five days a week (approximately 170 miles per working day). As well as passengers, the vehicle also carries pre-prepared meals and internal post. A desire to reduce environmental and congestion problems initially led the Trust to investigate the benefits of a cleaner minibus service. Oxford City Council had adopted its own green transport policy that includes the use of clean fuel vehicles, and the Trust is keen to meet these same objectives.

The minibus has a standard vehicle specification, and did not require any special modifications apart from the LPG conversion. It cost £14 000, with the LPG conversion costing an additional £1700. However, this was offset by a grant of £1000. The LPG vehicle would be expected to reduce by half the carbon monoxide, hydrocarbons, and oxides of nitrogen emissions of a comparable petrol vehicle operating in the streets of Oxford.

The minibus has proven to be very suitable for its specific operation, but the Trust is convinced it could be used for other duties if so required. The vehicle's performance has been satisfactory and meets the needs and requirements established in the Trust's Transport Plan. The transport manager also considers that: *'Considering the high mileage and punishing duty cycles we subject the vehicle to, it has proven very suitable for its use.'*

The Trust found that it was much cheaper to install LPG refuelling facilities at its main site, than to refuel at existing local alternatives. The LPG tank was supplied and installed on-site and no additional infrastructure was required. The LPG fuel cost is very competitive and there is an additional nominal rental fee for the gas storage tank. Significant cost savings over a conventional diesel vehicle have been achieved. Petrol, when needed, can be purchased locally.

Overall vehicle reliability has been very good.

In addition to the LPG minibus, the Trust also operates an LPG van used for general maintenance duties. There are several vehicles operated by the Trust that will soon need to be replaced. LPG will be seriously investigated as an alternative to the [2001] diesel options. According to one transport manager, 'Now that mainstream automotive manufacturers are producing bi-fuel, as well as electric vehicles, things should change.' The Trust remains firmly behind its decision to invest in cleaner-fuelled vehicles (CFVs).

Key facts

Featured vehicle	Bi-fuel Minibus
Conversion cost	£1700
PowerShift grant	£1000 [the PowerShift scheme is now discontinued]
Average monthly mileage	3000 miles
LPG fuel cost	[27p] per litre (bulk purchase) [2006 prices]
Economy (approx.)	20% less mpg of fuel compared to conventional vehicle
Emissions	Significant reductions in particles and oxides of nitrogen
Performance	Unchanged from petrol vehicle
Passenger numbers	11 people

Adapted from EST, 2001

2.8 Biofuels

Liquid biofuels are produced by the **fermentation** of energy crops or the **esterification** of vegetable oils or animal fats. These fuels can reduce the transport sector's dependence on fossil fuels and, in principle, their use can provide reductions in some regulated and greenhouse gas emissions on a life cycle basis.

Ethanol (CH_3CH_2OH; also known as **ethyl** or **grain** alcohol) is a clear, colourless liquid and is the essential ingredient of all alcoholic drinks. It can be produced from virtually any fermentable source of sugar. Ethanol made from cellulosic biomass materials instead of traditional feedstocks is called **bio-ethanol**. The production method first uses enzyme **amylases** to convert the feedstock into fermentable sugars (dextrose). Yeast is then added to the **mash** to ferment the sugars to ethanol and carbon dioxide.

Figure 2.14 A substantial proportion of vehicles in Brazil are fuelled by ethanol derived from sugar cane

Another alcohol fuel is methanol (CH_3OH; also known as **wood** alcohol), which is predominantly produced via **steam reforming** of natural gas to produce **syngas** (a mixture of carbon monoxide and hydrogen). This is then fed into another reactor vessel under high temperatures and pressures, where the gases are combined in the presence of a catalyst to produce methanol and water. Although over 80% of methanol is currently produced in this way, the ability to produce **bio-methanol** from non-petroleum feedstocks (including biomass) is of interest for reducing reliance on fossil fuels.

Bio-diesel is most commonly produced by the **esterification** of energy crops such as oilseed rape (OSR) or recycled vegetable oils (RVO). Animal oils can also be used (see Box 2.10). The oils are filtered and pre-processed to remove water and contaminants and are then mixed with an alcohol (usually methanol) and a catalyst. The oil molecules (triglycerides) are broken apart and reformed into fatty acid methyl esters and glycerol, which are then separated from each other and purified. Bio-diesel from OSR is known as rape methyl ester (RME). The fuel can be used (pure or as a blend) in place of mineral diesel in many modern diesel-powered vehicles. The production of RME also has two valuable by-products: glycerine, which is used in pharmaceuticals and cosmetics, and cattle cake made from the remaining plant material.

BOX 2.10 Chicken fat to power supermarket lorries

Starting in January [2003], Asda trucks of up to 40 tonnes will carry startling slogans saying 'This vehicle is powered by chicken fat' – the biggest boost yet for the legal use of recycled cooking oil on Britain's roads. Lorries making deliveries on Tyneside and in Yorkshire will be the first to try the fuel, which is currently available on three forecourts in Yorkshire. A further eight garages in the region are to take supplies from the growing number of biodiesel refiners, who were given a 20p-a-litre green tax concession by the Chancellor [2002].

Asda [a UK supermarket chain] produces more than 50m litres of used cooking oil and 138 000 of waste frying fat every year from its canteens, restaurants and rotisseries. The gunge was a disposal headache rather than a potential money-earner until an unexpected phone call last spring. 'We were approached by a biodiesel firm, which cleans up waste cooking oil, adds a bit of methanol and sells it as a much cheaper alternative to diesel,' said Rachel Fellows of Asda yesterday. 'We were only too happy to do business with them. But then we thought: hang on, isn't there something we can do here for ourselves?'

Company trials of 'chip pan fuel' for Asda's cars and lorries were then intensified after the firm's innocent involvement last month in a moonshine operation at Llanelli in South Wales. A special 'frying squad' set up by Dyfed Powys police discovered that hundreds of drivers were running their cars on Asda's 'extra-value' cooking oil mixed with methanol at home, in a moonshine operation which dodged tax. The 32p-a-litre fuel supply ... was cut off when Asda discovered its Llanelli branch was selling vastly more oil than anywhere else in the country. Rationing was imposed and the police frying squad – whose tactics included sniffing out the chip-shop smell of bootleg cars – moved in.

The planned Asda fleet fuel, like all commercial biodiesel, is completely legal but will still undercut conventional diesel prices by at least 10p a litre. Converting an in-house product like the waste oil will add to savings for the firm. 'Oil's a finite resource and we are fully aware of the fact that we shouldn't be wasting it,' Ms Fellows said. 'This is real eco-innovation – trials already show that chip pan fuel emissions are up to 40% lower than diesel.'

Wainwright, 2002

Vehicle technology

Because they are liquids at room temperature, ethanol and methanol can be handled in a similar way to conventional fuels. Both have high octane ratings (enabling a high engine compression ratio which increases engine efficiency). They can be used in spark-ignition (petrol) engines with little or no modification as alcohol-petrol blends (e.g. E10 is 10% ethanol; also known as gasohol) or as pure alcohol fuels in modified vehicles. The suitability of alcohols as vehicle fuels is demonstrated by their use as high-performance motor-racing fuels, for example in the Indianapolis 500.

The principal difficulty with alcohol fuels is their relatively low energy density. This means that vehicles running on pure alcohol require a storage vessel double the volume of an equivalent petrol tank. Also, as alcohols are difficult to vaporise at low temperatures, pure-alcohol vehicles are difficult to start in cold weather. For this reason, alcohol fuels are usually blended with a small amount of petrol to improve ignition. Methanol has the added disadvantage that it is both highly toxic and hydrophilic (mixes readily with water in all proportions), which can be a danger if used near to sources of potable water.

Alcohol fuels are already added to petrol to improve octane ratings and as **oxygenate** additives (to reduce carbon monoxide emissions). In the USA alone, more than 1.4 billion gallons of ethanol are added to petrol each year (ACE, 2006). Although many other countries, including the UK, also use ethanol as a petrol additive, most consumers of petrol vehicles are unaware that this practice occurs. Methanol reacted with isobutylene to form methyl tertiary-butyl ether (MTBE) is also used as an oxygenate additive. However, the use of MTBE as a petrol additive is being phased out owing to new health concerns associated with its use.

Bio-diesel is primarily used by heavy-duty vehicles, as this sector is almost wholly dependent on diesel engine technology and very few alternatives exist for trucks of high tonnage. Most modern heavy-duty diesel engines can use bio-diesel without modification provided the fuel is of the correct specification. Bio-diesel can also be blended in any proportion with mineral diesel. One potential problem of 100% bio-diesel fuel (B100) is an increase in the corrosion of rubber products. Engines and equipment with rubber seals and piping are usually replaced with non-rubber alternatives (a B5 blend does not lead to this problem).

As bio-diesel has a lower energy density than mineral diesel, its use results in an increase in fuel consumption of around 5% (a B10 blend would result in a 0.5% difference). Existing fuel tanks therefore give slightly less mileage when using bio-diesel. Another minor problem is that B100 is more viscous than mineral diesel in cold weather. However, a cold filter plugging point (CFPP) additive can alleviate this problem, enabling even pure bio-diesel to be used in temperatures as low as –22 °C.

Fuel supply and infrastructure

Ethanol is one of the most widely used alternative vehicle fuels in the world, due largely to its widespread use in Brazil and the USA. Over

90 production plants in North America are in operation providing fuel ethanol production from starch crops (primarily corn). About 35 million tonnes of corn are used annually to provide more than 4 billion gallons of ethanol for E10 alcohol blends, equivalent to around 3% of the US petrol market (ACE, 2006).

During the 1970s and 1980s, ethanol produced from sugar cane was vigorously promoted in Brazil both as a response to a slump in the global price of sugar and to reduce the country's dependence on foreign oil imports. At that time, all light-duty vehicles were required to run at least in part on ethanol fuel. In 1989, the country's total fleet of cars and light-duty vans consisted of over 4 million pure-ethanol and 5 million gasohol vehicles (Johansson et al., 1994). From 1973–87, even though the country's total energy demand almost doubled, petrol use dropped from 12% to only 4% of the energy market, while ethanol production increased to 18%. As a result of this and new home-production of oil and natural gas, the country's dependence on oil imports reduced by almost half. Since 2004, annual Brazilian ethanol production has exceeded 4 billion (US gallons) and a new generation of 'flexfuel' vehicles (FFVs) has been developed that can run on a range of petrol–ethanol blends (RFA, Ethanol Production Statistics, 2006). To date there is no large-scale UK-based production of bio-ethanol; however, in 2007 the Somerset Biofuels Project is set to produce bio-ethanol specifically for a fleet of flexible fuelled bio-ethanol cars. The project is a partnership including the local council, Wessex Grain and Ford Motor Company working together to contribute to fuel and vehicle production.

Bio-diesel is widely produced in Austria, Germany, France, Italy and Sweden. Following favourable signals from the EU, around 40 production plants annually produce around 3 million tonnes of bio-diesel (EBB, 2006). Although domestic bio-diesel production could reduce oil imports and improve energy security, there is as yet no significant large-scale UK production of bio-diesel. However, this is changing as Argent Energy operate the UK's first large-scale bio-diesel plant in Scotland, which uses waste oils to produce over 50 million litres of fuel per year. This marks the increasing importance of biofuels in the UK – whereas biofuels accounted for only 0.05% of fuel sales in 2004, this had increased by a factor of six by the close of 2005, and is well on the way to meeting the government's proposed biofuel target of 5% of all road fuels by 2010 (Ecolane, 2006). In 2005, the UK government has reduced the fuel duty on bio-diesel by 20p/litre, indicating a high level of support for the fuel in the future.

Environmental impact

Although difficult to quantify, the consensus is that CO, HCs and particulates are reduced for M85 (85% methanol and 15% petrol), E85 blends and pure-alcohol fuels. Though alcohol-fuelled vehicles can emit less nitrogen oxides (NO_x) (as alcohol fuels burn at a lower temperature than petrol), in practice the compression ratio is often increased to improve engine efficiency, increasing combustion temperatures and offsetting any reduction in NO_x emissions. Unburned alcohols present in the exhaust gases of an alcohol-fuelled engine contribute less to tropospheric-ozone formation than do the volatile organic compounds present in petrol exhaust emissions.

Regulated emissions for bio-diesel are generally reduced when compared to standard diesel operation. For example, estimates based on a number of comparative tests suggest that particulate emissions are of the order of 10–15% lower than with ULSD. Bio-diesel's low sulphur content also allows the use of advanced emission control systems, which can further reduce particulates. However, without any emission control system, NO_x emissions can be increased for bio-diesel by 5–10%.

The great promise of biofuels is their *potential* to be carbon neutral, all the CO_2 emitted during processing and use of the fuel being balanced by the absorption from the atmosphere during the fuel crop's growth. However, in practice, unless organic growing methods are used, this is rarely the case, as the process of growing the biomass requires the input of fossil fuels for fertilisers, harvesting, crop processing and fuel distribution. The actual extent of total greenhouse gas emissions is therefore strongly dependent on the energy crop and the fuel processing used.

For example in Brazil, where sugar cane is used as the feedstock for ethanol production, large amounts of *bagasse* (woody fibres remaining after the juice is extracted from the cane) are used to provide the process heat energy. As a result, the average energy ratio of ethanol output to fossil fuel input is of the order of six, i.e. six units of energy are produced for each unit input. Therefore, on a life cycle basis, carbon emissions are significantly reduced by up to 90%. This contrasts with the net energy ratio for corn-derived ethanol from the USA which, in some cases, can be negative, i.e. the fossil fuel required to produce the ethanol is *greater* than the energy value of the final product.

The same wide variation in life cycle CO_2 emissions is true for methanol, as the emissions depend on the feedstock and processes employed. Although emissions associated with methanol from biomass can be lower than for conventional fuels, if fossil fuels are used as energy feedstocks there is little difference between methanol and using petrol or diesel. Similarly, the results of life cycle analysis of greenhouse gas emissions for bio-diesel depend on the production processes employed. However, studies show that, for RME, these emissions can be reduced by around 40%, even when upstream emissions from the production of fertiliser are included in the analysis (Concawe, 2005).

BOX 2.11 Advantages and disadvantages of vehicle biofuels

Advantages:

- **Reduced emissions –** can reduce life cycle greenhouse gas emissions by up to 90%
- **Security of fuel supply –** useful alternative to importing of crude oil products.

Disadvantages:

- **Engine modification –** vehicles switching from conventional to biofuels may require minor modifications or adjustments
- **Land requirements –** large amount of land area required to supply existing vehicle fleet.

2.9 Battery electric vehicles

Battery electric vehicles (BEVs) are ideally suited to applications that benefit from zero-emission operation. These include use as small city cars, light-duty vans for freight delivery and industrial vehicles such as forklift trucks that are used within an enclosed space. There are around 40 000 BEVs in Europe, which includes over 15 000 milk delivery vehicles in the UK, one of the largest BEV fleets in the world. However, other than milk floats, there are only around 300 modern electric vehicles in use on British roads (Avere, 2006).

Vehicle technology

The design principle of a BEV is relatively simple. Electrical energy (from any source of primary energy) can be stored in a 'secondary' or rechargeable cell on-board the vehicle. When required, electrical energy is drawn from the cells and converted to motive power by the use of an electric motor.

BEVs are significantly more energy-efficient than conventional vehicles in stop–start traffic, as they use almost no energy when 'idling'. Electric vehicles can also recover the energy usually lost when braking via regenerative braking systems (up to 20% can be recovered). For these reasons, a battery with a **specific energy density** (defined as the energy content per unit mass) of around 200 Wh kg^{-1} would provide a small BEV with a range comparable to that of a conventional passenger car.

The tried-and-tested lead–acid battery is the most widely used for BEVs. Although they have a relatively low specific energy density (30–40 Wh kg^{-1}), it is possible to build a vehicle that has a range of around 70–90 km using lead–acid technology. Although these cells are far from ideal in their energy storage and power delivery characteristics, they are known for their reliability and durability, and are supported by an extensive maintenance network.

Other common traction batteries include nickel–cadmium (Ni–Cd), nickel–metal–hydride (Ni–MH) and lithium–ion (Li–ION). Their higher energy density (50–90 Wh kg^{-1}) provides a significant improvement on lead–acid technology, increasing both vehicle performance and range. However, these battery types are expensive to produce and their use involves handling toxic materials such as cadmium and lithium. Indeed, in 2006 the EU banned the use of cadmium for battery production. Despite these problems, these Ni–MH and Li–ION batteries have proved to be well suited to motive applications and are now preferred by many BEV manufacturers. Any large uptake in battery-powered vehicles could result in some types becoming prohibitively expensive owing to the price/demand sensitivity of the component metal materials.

Most first-generation BEVs used **direct current** (DC) motors that are relatively cheap, give high torque at low speed and are easy to control using semiconductor technology. However, their efficiency of 80–85% and *specific power density* of 150–200 W kg^{-1} (about a third that of a petrol engine) does not represent the best possible performance of available motor technology. One alternative is to use the **alternating current** (AC) induction motor, which has increased efficiency and double the specific power, but involves the use of a more costly control system.

The main disadvantage of BEVs is their high capital cost (typically an increase of 50–100%). In addition, most BEVs do not match the performance of conventional vehicles. However, most BEVs have a range and performance which is adequate for many specific, urban applications and are particularly suited to drive cycles that are predictable, regular and less than 100 km per day (e.g. delivery cycles), especially in areas where low-emission vehicles are preferred or mandated. BEVs are therefore well-suited for use in commercial fleets (for small loads), company car pools and within rental fleets.

Fuel supply and infrastructure

The most common charging cycle is an overnight 'trickle-charge' from a standard domestic 13A, 230 V socket. This typically takes 6–8 hours and requires the use of a transformer to reduce the voltage and the current, which is then rectified to charge the cells using DC. Fast charging (which takes less than one hour) is also possible but requires the installation of specialised recharging points.

Using the national electricity grid, it is relatively easy and inexpensive to install slow-charge points as compared with other alternative fuels. Where an existing socket is not available, total installation costs per (standard) charge point are in the order of £500. However, fast charging systems (required for publicly accessible refuelling points) cost in the order of £7000–£30 000 per point (depending on type).

Most charging systems use a conductive cable to transfer the electrical energy to the vehicle. However, this is not the only option. **Inductive** (non-contact) charging systems have also been successfully demonstrated in the French Praxitèle project. Yet another approach to vehicle recharging is to recharge the batteries away from the vehicle. In Birkenhead, UK, six Techobuses use battery packs that are recharged at the fleet's base. When refuelling is required, the depleted battery pack is exchanged for one that is fully charged.

BEVs have low fuel costs per km, due to the low price of electricity relative to other road fuels, and to the high efficiency of the electric drive-train. For example, a typical battery electric car costs less than 1p/mile to run (compared to a *fuel* cost of around 8p/mile for a petrol car) (Ecolane, 2006). Over an average annual mileage of 19 000 miles per year, this represents a cost saving of around £700 per year. However, if battery lease costs are taken into account (about £70 per month), this saving is negated. It is therefore very difficult to accumulate the mileage necessary to recoup the extra capital required. Costs, therefore, remain a significant barrier to the introduction of BEVs.

Environmental impact

The battery electric vehicle is essentially a zero-emission vehicle *at the point of use*. Electricity used to recharge BEVs can be generated by the combustion of primary fossil fuels, the fission of nuclear fuels or can be produced using renewable sources. If renewables are used, a BEV can be operated with zero fuel-associated emissions on a life cycle basis.

As electricity is produced from a range of energy sources (including coal, nuclear, oil, hydro, natural gas, and increasingly wind, solar and wave power), we need to consider the production processes in detail if we are to be able to analyse the impacts of electricity use within the transport sector. Using a typical UK fuel mix, the *life cycle* data shows that CO and HCs are reduced for BEVs (as compared to petrol), although with an increase in particulate and sulphur emissions. For life cycle CO_2 emissions of grid-electric-fuelled BEVs, data shows a reduction in greenhouse gas emissions of approximately 45% (compared to a petrol baseline). The trend is towards a generally cleaner electricity generating mix, with an increased fraction of combined cycle gas turbine (CCGT) plant and renewables. Indeed, if electricity from renewable sources is used, the fuel life cycle emissions will be very small.

The benefits of BEVs to urban air quality are twofold: lowering the overall emissions (gaseous and noise), and removing the emission sources from urban areas where the greatest number of people work and live. Furthermore, predicted life cycle emissions reductions are often underestimated, as the equivalent emissions for other fuels are based on hot engine conditions and do not account for cold start conditions when a high proportion of emissions from ICE vehicles can occur. However, the increased quantities of heavy metals (e.g. lead and cadmium) in circulation due to the increased uptake of battery-powered vehicles would require addressing from a life cycle perspective.

BOX 2.12 Advantages and disadvantages of battery electric vehicles

Advantages:

- **Zero-emission at point of use** – can also utilise renewable electricity, so providing life cycle zero-emission transport
- **Reduced noise** – BEVs are almost silent at slow speeds and have low vibration in operation
- **High efficiency** – electric drive systems are more energy-efficient than ICEs in stop–start driving
- **Regenerative braking** – can recover up to 20% kinetic energy normally 'lost' in a conventional vehicle.

Disadvantages:

- **High capital vehicle cost** – high cost of electric drive-train and batteries can double the vehicle capital cost
- **Limited vehicle range** – typical small BEV has a range of less than 100 km owing to limitations of battery energy storage
- **Long recharge time** – typically 6–8 hours for a slow charge
- **Increased vehicle weight** – battery pack increases vehicle mass by 300–900 kg.

BOX 2.13 The G-Wiz automatic electric vehicle

Figure 2.15 The tiny footprint of the electric car allows parking in places otherwise unused

There is a map on the wall of the office of GoinGreen's offices... [the company that markets the G-Wiz battery electric car; see Figure 2.15)], which shows the spread of emission-free motoring. It looks like the early stages of a virus, with coloured pins marking the address of every owner of a Reva G-Wiz [shown in Figure 2.15].

So far, the map is restricted to Greater London. The armies of pins have outposts as far as Chislehurst and Beckenham in the south-east, and Wimbledon in the south-west, stretching as far north as Barnet. There are a couple in Ealing. The big battalions are clustered in the leafier parts of north London, with high concentrations of colour in Primrose Hill and Hampstead. Since they went on sale in summer 2004, more than 500 of these impish electric cars have been sold.

The analogy with a virus is apt. GoinGreen doesn't advertise its cars, instead selling them by word of mouth and through its website: www.goingreen.co.uk. If the pins clump together, it's because the owners tend to recommend them to their friends.

Although available in the UK, the G-Wiz is built in Bangalore, and was conceived in California by Dr Lon Bell, an engineer who made his fortune making airbag sensors and seatbelt tensioners, before becoming intrigued by the way cars work. In designing an electric car, he decided to ignore the assumptions of conventional construction. His first thought was to ask what was necessary in a car, from which he concluded that it needed wheels, with tyres, something to steer and a windscreen. Most of the rest was luxury and got in the way of making a nimble, no-frills electric vehicle for non-polluting urban travel.

The G-Wiz is designed to seat two adults and two small children. Although smaller than conventional cars, there being no engine as such, both bonnet and boot can be used for storage. As well as a body made from dent-resistant plastic, it has regenerative brakes: pressing the pedal works like a dynamo, recharging the engine. It also has climate-controlled seating. Each seat has tiny heat-releasing holes which warm the body rather than the air in the car. There is a conventional heater, too, but using it will knock 10 miles from the car's 40-mile range.

To the non-mechanically-minded, the G-Wiz is a remarkable piece of technology. It requires only a little more attention than a mobile phone. To charge it, you stick a lead in the socket where the petrol cap should be, and you have to water the battery every two or three weeks. 'It's like a plant,' says Joe Byers of GoinGreen. 'Every so often a light will come on saying "Water me, please".'

This procedure is simpler than topping up a steam iron. You don't have to open the bonnet. You stick a small pipe into a hole by the plug, hold it in the air and pour in distilled water. You need never touch an oil can. Oiling is done during servicing. 'You water your car,' says Joe. 'That's all you need to do.'

Driving the thing is marginally more complicated, but will not test the aptitude of anyone who has ever sat in a dodgem. There are two pedals – an accelerator and a brake. The handbrake is a twisty device under the dashboard on the left of the steering wheel. The ignition is on the right. This is the first big shock. When you turn on the engine, nothing happens. Actually, that's the point. There is no engine. When you turn the key you are not greeted by an angry growl of machinery. There is nothing, unless you count the flickering of a small green light on the dashboard. At first, this is disconcerting. Without the engine noise, the instincts of conventional driving don't kick in.

There is no pumping of the accelerator or gentle easing of the brake, and none of the sense of power which is at the root of all car advertising. In this little moment of uncertainty, with no rush of testosterone to the places that make urban motoring slightly less relaxing than bare-knuckle boxing, it's tempting to forget the routines of driving – the mirror, signal, manoeuvre bit. Fortunately, such disorientation is not that dangerous. The G-Wiz seems to think before it moves, and when it does, it's a stately glide.

There are no gears. The car has a dial, with four modes: reverse, neutral, economy, and full power. In London, where the average speed of travel is less than 10 mph, full power (with a top speed of 42 mph) is rarely necessary, but it does offer slightly more 'oomph' when easing from traffic lights.

On the open road, there is a perplexing absence of noise. Suddenly, you are aware of the volume from other cars' engines. Aurally, it's a bit like being a non-smoker in a cigar bar: you find yourself defined by the thing you are not doing. But it does make you wonder how much quieter our cities would be if all the short journeys were electric.

The car is cute ... and its green credentials are impeccable. But it is economically attractive, too. The G-Wiz is exempt from road tax, as it produces no carbon emissions. Since it costs only 40p to charge the car for 40 miles of driving, GoinGreen calculates that a London commuter will save the cost of the car (£7799) in [two to three years, as the G-Wiz is eligible for the full London Congestion Charge discount and can park free in many London Boroughs].

The biggest problem for the spread of the technology is the need for off-street parking during the recharging process. Some London car parks offer recharging facilities, but flat-dwellers or owners without a driveway will need support from councils to make the G-Wiz a practical option. Similarly, potential drivers outside London will have to wait until the company expands, or the technology becomes more universal, as servicing is currently only available at GoinGreen's headquarters. How does it feel? Well, not sexy exactly, but there is something endearing about the car that seems to bring out the best in other road users.

[...]

Lon Bell has compared the G-Wiz to the early mobile phones. In later models the batteries will be smaller and more efficient. A prototype of a hard-top roadster already exists: [it can travel at up to] 80 mph and has a range of 100 miles. That may bring more torque to the electric revolution, but it will be hard-pushed to replicate the ... charm of the little G-Wiz [see Figure 2.16].

Adapted from McKay, 2006

Figure 2.16 The battery electric G-Wiz is exempt from the London Congestion Charge and is provided with free parking and dedicated recharging points in many parts of Central London

2.10 Fuel cell electric vehicles

> There seems to be a feeling creeping through the motor industry that perhaps the days of the internal combustion engine are numbered
>
> Hart and Bauen, 1998, p. 7

There are many reasons to support the transition from a carbon-based energy system to a **hydrogen economy** (Boyle et al., 2003). Primarily, this is to reduce overall carbon emissions, which are associated with climate change. However, the use of hydrogen as a fuel also provides other advantages. For example, hydrogen gas has the highest energy-to-weight ratio of all fuels, with 1 kg of hydrogen containing the same amount of energy as 2.5 kg of natural gas or 2.7 kg of petrol (see Tables 2.1 and 2.5).

The use of hydrogen as a fuel is not new. 'Coal gas' or 'town gas', which is at least 50% hydrogen, has been used extensively throughout the industrial nations and preceded the use of natural gas in North America and Europe. Around 1.5% of world energy supplies are already converted to hydrogen gas for use in the chemical and petrochemical industries (Boyle et al., 2003). The gas is typically used for the chemical synthesis of ammonia, ethylene and methanol and in the desulphurisation and hydrogenation of fossil fuels.

Hydrogen is a versatile fuel and can be used within modified internal combustion engines (Boyle et al., 2003). Since the 1970s, BMW has developed a series of hydrogen ICE prototype cars, the latest of which include the bi-fuel 745hL and the hydrogen-powered MINI. The only combustion products from a hydrogen-powered ICE are water vapour and small amounts of NO_x (due to the presence of atmospheric nitrogen). In addition to reduced emissions if used within an ICE, the use of hydrogen as a fuel also offers the possibility of using an alternative engine technology, the **fuel cell**.

The fuel cell

Fuel cells are electrochemical devices that convert chemical energy directly into electrical energy, heat, and water. The principles of fuel cells are similar to those of electric batteries, where energy conversion takes place between the reactants to produce electricity (Box 2.14). However, unlike a battery, a fuel cell does not store chemical energy. The reactants (fuel and oxidant) have to be continually supplied to the cell for an electric current to be produced.

Most fuel cells consist of two electrodes, an 'anode' and a 'cathode', which can be made from a variety of electrically conducting materials. The anode and cathode are separated by an **electrolyte** that allows the transfer of ions, but physically separates the fuel and oxidant. This prevents the exchange of electrons that would be required for a non-catalytic chemical reaction to occur (i.e. combustion). When the reactants are fed into the cell, chemical reactions occur between the fuel/oxidant and the electrolyte. The main charge carriers (usually H^+) cross the electrolyte and the electrons are transferred via an external circuit. The electric current produced can be used to drive a motor or other external load. The fuel normally used is hydrogen or a hydrogen-rich compound (supplied to the anode) and the oxidant can either be pure oxygen or air (supplied to the cathode)(Figure 2.17).

Many people will be familiar with school experiments in which water is split into its constituents, hydrogen and oxygen, by the process of 'electrolysis' – passing an electric current between two electrodes immersed in water. Fuel cells operate in a manner that is essentially the reverse of electrolysis – by combining, rather than splitting, hydrogen and oxygen. This process generates an electric current, water – and some 'waste' heat. Fuel cells can use any two reactants that are respectively oxidising (i.e. a source of oxygen) and reducing (i.e. readily combine with oxygen), but the most common reactants are hydrogen and oxygen (or air, which is approximately 20% oxygen).

The anode and cathode are usually coated with platinum, or a platinum-group metal such as palladium or ruthenium, which acts as a catalyst. Catalysts are substances that increase the rate of a chemical reaction without themselves undergoing any permanent chemical change. Between these is placed an 'electrolyte', of which again there are a variety of types. Normally hydrogen is the fuel fed to the anode, while oxygen (from air) is supplied to the cathode. Both the anode and the cathode are porous, allowing the gases to flow through them. With the aid of the catalysts present on the surface of the electrodes, the hydrogen splits into hydrogen ions (i.e. protons) and electrons. The electrons flow away from the anode into an external electrical circuit where they can be made to deliver useful energy. Meanwhile, the hydrogen ions flow through the electrolyte to the cathode, where (again with the aid of a catalyst) they combine with the oxygen supplied to the cathode and the incoming electrons from the external electrical circuit to form water vapour. Depending on the type of cell, typically 30–60% of the energy content of the input fuel is converted to electricity – the rest appears as heat, but this can often be used, either for space or water heating or to provide energy for the 'reformers' that may be required to convert, say, natural gas into the pure hydrogen required by the fuel cell.

The key advantage of the fuel cell as an energy converter is that electricity is produced *directly*. This means that its efficiency can be higher than the limits set by Carnot for heat engines. It also means that there are no emissions of the gaseous pollutants that are associated with combustion processes, such as SO_2, NO_x or particulates. If pure hydrogen is the fuel, there are no CO_2 emissions and the only other emission, apart from some 'waste' heat, is water vapour.

Single cells typically generate around 0.8 V with a power output of up to 100 W. Larger outputs are achieved by assembling cells in series or parallel to form a **stack**, which has the required voltage and output characteristics. In contrast to heat engines, fuel cells are not limited by the Second Law of Thermodynamics, which means that they are able to achieve higher conversion efficiencies than heat engines (Boyle et al., 2003). Although there are losses within a fuel cell that arise due to ohmic resistance of the cell components, efficiencies of up to 80% have been demonstrated in the laboratory.

BOX 2.14 Principle of operation of a PEM fuel cell

The PEM fuel cell uses highly conducting electrodes made of graphite, which form the terminal of each cell and separate adjacent cells in the stack. The electrodes are grooved to allow easy passage of the gases to the 'surface of action' while also maintaining electrical contact with the electrolyte-catalyst-gas interface. At the anode, hydrogen is catalytically disassociated to leave hydrogen ions. An external circuit conducts electrons, while the positive ions (protons) migrate through the electrolytic membrane to the cathode. There they combine, again under action of a catalyst, with oxygen and electrons returning from the external circuit, to form water.

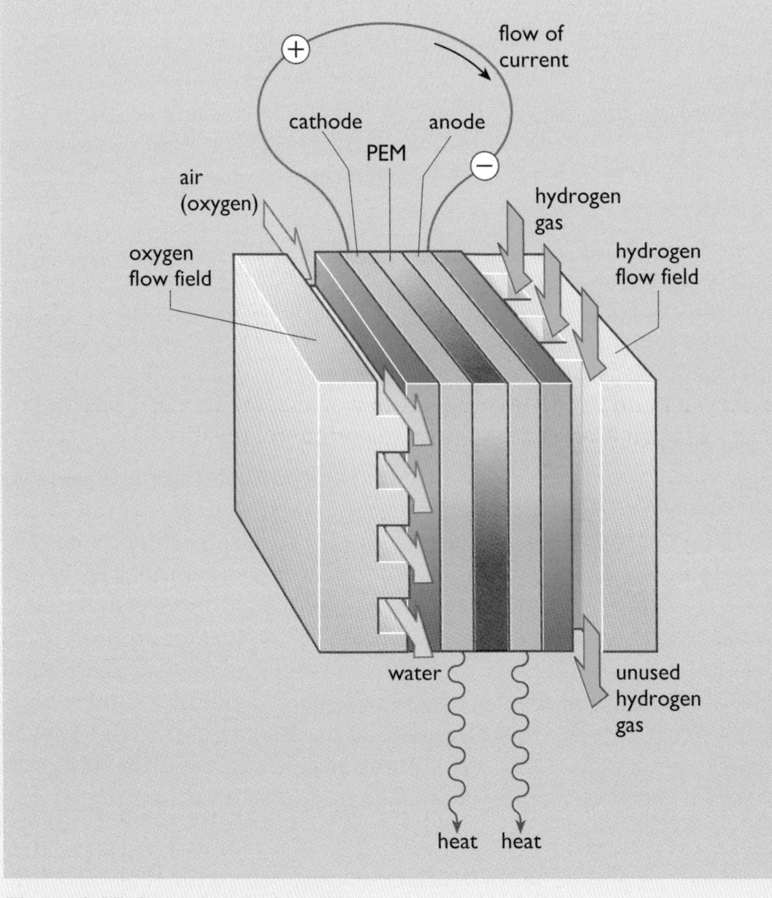

Figure 2.17 Principles of a fuel cell

Several fuel cell types have been developed, each being characterised by the electrolyte used, operating temperature and fuel gas quality required. Low-temperature fuel cells (approximately 70–90 °C) include the alkaline fuel cell, the solid polymer fuel cell, of which there are two types: the proton exchange membrane fuel cell, and the direct methanol fuel cell. Owing to the relatively low temperatures within the cells, these usually require a catalyst at the anode to promote the necessary reactions taking place. High-temperature fuel cells (650–1000 °C) include the molten carbonate fuel cell and the solid oxide fuel cell.

Vehicle technology

If fuel cells are to replace the internal combustion engine in road vehicles, they need to have comparable power and a similar response time. In practice, this means a power density of at least 1 kW kg^{-1} (for cars) and a start-up time measured in seconds. The fuel cell thought by most analysts to meet these requirements is the PEM fuel cell, which has the ability to operate at relatively low temperatures, so reducing start-up times. Solid polymer electrolyte materials such as Nafion®(related to Teflon®) also eliminate the safety considerations associated with liquid acid and alkali electrolyte cells.

Following the successful use of fuel cells in the Gemini and Apollo space missions (which used solid polymer and alkaline fuel cells respectively), the 1960s saw a number of terrestrial fuel cell vehicle prototypes. These included Shell's 20 kW fuel cell truck and General Motor's liquefied hydrogen-oxygen-fuelled Electrovan, which was powered by a 5 kW Union Carbide fuel cell. In the 1970s, interest in fuel cells was renewed owing to the sharp increase in world oil prices and the decade saw designs such as the hybrid AFC-battery Austin A40 car, which used roof-mounted compressed hydrogen tanks and had a range of 300 km (Hart and Bauen, 1998). However, interest in road fuel cell vehicles declined in the 1980s as the fuel crises of the 1970s receded.

In the 1990s interest in fuel cells for road transport was revived, this time with a focus on the environmental benefits that the technology could provide. This decade saw the development of **fuel cell vehicle (FCV)** prototypes by most of the major vehicle manufacturers and the emergence of new companies specialising in the manufacture of fuel cell systems. One such company is Ballard Power Systems who, in collaboration with DaimlerChrysler and Ford, developed the world's first fuel cell bus and the Necar (New Electric Car), with a view to commercialisation.

The Necar programme was specifically designed to develop a commercial PEM fuel cell vehicle. From 1994 to 2000 five prototypes were tested (Necars 1 to 5), with the objective of reducing the mass and volume of the fuel cell stack and on-board fuel system to a size suitable for passenger car applications. During the programme, the stack power output improved from 5 kW to 75 kW. Three on-board fuel systems were also tested; Necars 1 and 2 used compressed hydrogen, Necar 4 was fuelled with liquid hydrogen and Necars 3 and 5 used methanol and an on-board **reformer** to generate hydrogen on demand (see next section). The Necar 5 achieved a top speed of 150 km h^{-1} and a range of 400 km on an 11-gallon tank of methanol (HyWeb, 2001).

The General Motors concept car, called the Hy-Wire, is a radical design in car body shape as well as in its underpinnings. The car employs third-generation fuel cell power systems, but along with them uses drive-by-wire technology, hence the name Hy-Wire (see Figure 2.18). This results in a vehicle which has far fewer traditional mechanical linkages. For example the car has no

Figure 2.18 The General Motors concept car, Hy-Wire, is based on a platform module which contains a hydrogen-powered fuel cell along with complete replacement of traditional mechanical components by using drive-by-wire systems.

combustion engine, no instrument panel and no foot pedals, as they have been replaced by the fuel cell power system and an advanced electronic control unit. By consolidating the fuel cell stack and the majority of the components into the chassis slab, which lies just below the floor level, it is also possible to change the body shape significantly. This type of body on a frame represents a new type of architecture for vehicles and may greatly help the introduction of fuel cell systems by reducing the economy of scale required to bring such a new production process to the end consumer.

Sixth-generation vehicles have been developed: the 'F-Cell' FV is part of the California fuel cell partnership demonstration programme, and it is anticipated that these will have superior performance. It is also clear that the US 'Freedom Car Targets' are very ambitious and are partly based on the fact that any new hydrogen-based vehicle will have to outperform the incumbent gasoline-based vehicle in terms of fuel cost, miles driven per tank of fuel and overall production cost of the vehicle, especially from the consumer perspective of purchase price of the car. These targets are outlined in Table 2.6 and although costs of fossil fuels may have risen since the targets were first set, they are still useful in terms of comparing new technologies with incipient ones. Clearly the targets are difficult, given the relatively short time frame for achievement and the scale of the task for a major infrastructure change. By basing the targets on full functionality of a typical passenger car, the programme is attempting to ensure consumer acceptance if and when the new technologies come to market.

Table 2.6 Some key technical targets for future hydrogen fuel and vehicles

Parameter	Goal (year)
Highly developed commercial codes and safety standards for H_2 fuel	(2015) to be widely demonstrated
Target distance driven per tank	> 300 miles (per tank of fuel)
H_2 fuel cost, regardless of production pathway	\$2–3 per gallon gasoline equivalent as a consumer price (2015) with widespread availability
Specific energy density per vehicle target	3.0 kWh kg^{-1} (9% wt. H_2) and 2.7 kWh $litre^{-1}$, and \$2 kWh^{-1} (2015)
Cost target of power train system	\$45 kW^{-1} (2010) and \$30 kW hr^{-1} (2015)

Sources: US Department of Energy, 2006a, 2006b and 2006c

DaimlerChrysler have also developed the Nebus, a fuel cell version of a 70-passenger single-deck bus. High-pressure (30 MPa) hydrogen cylinders mounted on the roof were used to power ten 25 kW PEM fuel cell stacks providing a range of 250 km. The Nebus was initially tested in Germany and North America (in Chicago and Vancouver). These successful demonstrations were followed by a development of the ZEbus (Zero Emission Bus), which was used as part of the California Fuel Cell Partnership programme. Further trials include the Clean Urban Transport for Europe (CUTE) fuel cell bus programme, which is demonstrating 30 fuel cell 'Citaro' buses (see Box 2.15) in several European countries, including the UK (London).

Box 2.15 **CUTE**

CUTE (Clean Urban Transport for Europe) is a major demonstration project which is co-financed by the European Union. It consists of nine cities in Europe as well as another two cities associated with ECTOS (Ecological City Transport System) in Iceland and STEP (Sustainable Transport Energy for Perth) in Australia. Each city tested three buses, called the Fuel Cell Citaro, and specifics for that model are listed below. The buses were to be run over a period of 2–3 years to collect technical data for the bus and Ballard, the fuel cell manufacturers, but in many cases the demonstrations of technology have been extended further.

Each city will be testing and demonstrating the first generation of fuel cell buses under a wide variety of conditions in order to collect data on different production and usage pathways. Both buses and stations will be examined in detail for lessons learned about which sub-components are most applicable and most efficient overall. Safety and public awareness and acceptability were all key issues of consideration in the design of the entire programme. In some cases these were adapted for city-specific requirements. Some of the general station specifications are listed here.

CUTE fuel station requirements included:

- small spatial footprints
- full service and support from the suppliers
- components which could be incorporated into any previously existing station design
- simple and rapid refuelling procedures and processes
- correct hydrogen quality
- the ability to produce hydrogen (if made on site) at part load.

CUTE, 2004

Fuel Cell 'Citaro' characteristics based on the Stockholm buses (Haraldsson et al., 2005):

- Approximately 12 metre length production series made by EvoBus
- Body reinforced to handle additional weight due to hydrogen tanks and associated body work
- Gross weight: 18 000 kg
- Passenger load: 57 (maximum, 32 seated)
- Power of fuel cell stack: 150 kW each (total of two stacks delivering 250 kW)

- Main electrical motor: 205 kW
- Maximum speed: 80 km h^{-1} (speed limiter installed)
- Hydrogen storage: 1845 litres (in nine tanks), or 40 kg H_2 (15 °C, 350 bar)
- Maximum pressure per cylinder: 350 bar
- Typical fuel consumption: 2.2–2.5 kg H_2/10 km driven.

Although the buses in general were designed for high reliability, robustness and low maintenance, the next generation of fuel cell buses are likely to have higher efficiencies and benefit from regenerative braking and hybridisation in order to save even more fuel. Continued work on heat rejection will also help fuel efficiency along with greater integration of components.

Feedback from passengers has been generally very favourable, but in the case of Stockholm's residents, 64% of passengers have said they would not be willing to pay higher fares in order to fund fuel cell technology on the buses (Haraldsson, et al., 2006). It is clear that for the acceptance of hydrogen much work is still needed if the pro-hydrogen contingent wants to help shape consumer attitudes. Initial work in this area suggests that education, marketing and exposure to the product have the largest roles to play in achieving this (Schulte et al., 2004).

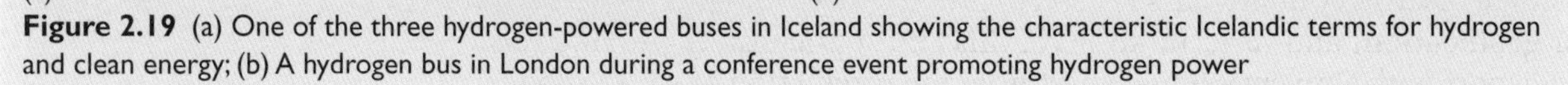
(a) (b)

Figure 2.19 (a) One of the three hydrogen-powered buses in Iceland showing the characteristic Icelandic terms for hydrogen and clean energy; (b) A hydrogen bus in London during a conference event promoting hydrogen power

Hydrogen's low density has presented a technological challenge to the design of on-board hydrogen storage systems. At room temperature and pressure, to store an equivalent amount of energy to that contained in a typical petrol tank would require a hydrogen tank with around 800 times the volume. From a technical perspective, three main methods of on-board hydrogen storage are currently under consideration. These are **compressed gas**, **liquefied gas** and **metal-hydride** storage.

Compression is the least expensive of the three options, the gas being stored in cylinders at pressures up to 30 MPa (300 bar). The most advanced tanks, which incorporate lightweight materials such as aluminium and carbon fibre, can achieve up to 3.6 MJ kg^{-1} (this compares with 32 MJ kg^{-1} for petrol plus tank). Cryogenic systems, which store hydrogen in its liquid state at low temperature (–253 °C), can achieve an energy density of around 16 MJ kg^{-1}.

However, ultra-low-temperature systems are expensive and liquefaction requires large amounts of energy, the energy required being about 40% of the energy stored. The third storage option is to use metal-hydrides that absorb hydrogen when under pressure, the gas becoming part of the metal's physical structure. The advantages of hydrides are the low loading pressures (less than 10 MPa), ease of use and high level of safety. However, hydrides are limited by their low energy density (up to 1.4 MJ kg^{-1}) and the complexities of the refuelling equipment (HyWeb, 2001).

Other methods of hydrogen storage being developed include the use of **carbon adsorption**, whereby hydrogen is adsorbed by carbon nano- or micro-fibres under pressure. Initial evidence suggests that this technique could enable storage densities higher than are achieved with liquefaction. However, these technologies remain at the development stage (Bérnard and Chahine, 2001).

Fuel supply and infrastructure

While there is a high level of agreement regarding which type of fuel cell is most suitable for road transport applications, the same cannot be said for the fuel supply system. This is primarily because of the large number of energy conversion routes that can be used to deliver hydrogen to the fuel cell. (Note that, like electricity, hydrogen is a secondary fuel or 'energy carrier' and must therefore be produced from primary energy sources.) However, the large number of production routes is also one of the great strengths of the hydrogen economy as the gas can be produced from almost any primary energy source.

Large-scale processes developed for the production of hydrogen from fossil fuels are well established. Currently, over 80% of hydrogen production is sourced from natural gas using the process of steam reforming. Other carbonaceous feedstocks (such as methanol, ethanol and biomass) can also be used to generate hydrogen via processes that include **thermal decomposition**, **partial oxidation**, and **gasification**. Alternatively, water can be **electrolysed** to produce hydrogen, (though it should be noted that energy is required to 'split' H_2O into its constituent elements, and therefore water should not be considered as the 'fuel'). If renewable energy in the form of electricity or biomass is used to produce hydrogen (via **electrolysis** or **gasification**) and used in an FCV, this could potentially provide road transport with zero emissions (apart from the production of water vapour) on a life cycle basis.

Box 2.16 Hydrogen as a fuel cell fuel

Hydrogen has been widely advocated as an 'energy carrier' for the future.

Its use as a fuel has many advantages:

- it can act as a temporary store of renewable energy from season to season
- it can provide a transport fuel that is not dependent on the world's declining reserves of oil
- the only by-products of its combustion are water and a very small amount of nitrogen oxides, and even the emissions of these can be reduced to zero if fuel cells are used.

Hydrogen is already used in large quantities as a feedstock for the chemical industry, mainly in the manufacture of fertilisers. Currently, it is mainly produced by steam reforming of natural gas (methane) which produces hydrogen and carbon dioxide:

$$2H_2O \quad + \quad CH_4 \quad \rightarrow \quad CO_2 \quad + \quad 4H_2$$
$$\text{steam} + \text{methane} \rightarrow \text{carbon dioxide} + \text{hydrogen}$$

Methanol can also be steam reformed to produce hydrogen, with CO_2 as the by-product.

One possibility is a hydrogen economy using fossil-fuel sources together with **carbon capture and sequestration**. However, renewable or 'solar' hydrogen can be produced without CO_2 by-products, in a number of ways:

- by the **electrolysis** of water using electricity from non-fossil sources. If direct current electricity is passed between two electrodes immersed in water, hydrogen and oxygen can be collected at the electrodes. This process could be used to produce hydrogen from renewable electricity virtually anywhere: solar plants in the deserts of Africa, wind power in the north of Scotland, or geothermal energy or hydropower in Iceland.
- by the **gasification** of biomass. Large amounts of hydrogen can be produced leaving a residue of high-grade carbon for chemical purposes. This carbon is of course likely to end up as CO_2 , but will be re-absorbed as long as the biomass is sustainably grown.
- by the **thermal dissociation** of water into hydrogen and oxygen using concentrating solar collectors. To do this directly would require very high temperatures, over 2000 °C, but with more complex processes using extra chemical compounds the same result may be achievable at temperatures of under 700 °C. These processes have not yet been developed on a commercial scale.

Other techniques are under investigation, including the use of photoelectrochemical cells that produce hydrogen directly from water via artificial chemical **photosynthesis**.

Using hydrogen as a fuel is well understood. 'Town gas' produced from coal before the arrival of natural gas consisted mainly of a mixture of hydrogen and carbon monoxide. Space rocket motors also run on a mixture of liquid hydrogen and liquid oxygen.

When burned, 1 kg of hydrogen will produce 120 MJ of heat, assuming that the resulting water is released as vapour. Although this is nearly three times the energy per unit *mass* of petrol or diesel fuel, hydrogen has the disadvantage of being a gas, with a low energy per unit *volume* at atmospheric pressure. It can be stored in a number of forms:

- as a gas in pressurised containers, typically at around 300 atmospheres. These containers obviously have a weight penalty
- by absorbing it into various metals, where it reacts to form a metal 'hydride': the hydrogen can be released by heating
- as a liquid, although this requires reducing its temperature to –253 °C and the use of highly insulated storage. Natural gas (methane) is already widely shipped in liquid form, but this only requires temperatures of –62 °C.

The main energy conversions leading to hydrogen can be categorised according to the location of hydrogen production, which can occur in one of three ways. Firstly, hydrogen can be produced centrally and then distributed to fuel stations where it is compressed and stored ready for use by an FCV. As is the case with large-scale production methods, a great deal of experience has been accumulated regarding hydrogen distribution on an industrial scale. For example, hydrogen is routinely transported by road in compressed form using steel bottles at 20 MPa, liquefied hydrogen is carried using 5000-litre capacity road tankers, and hydrogen gas is also routinely piped under pressure. There are already almost 100 operating hydrogen vehicle refuelling stations worldwide, which include public-access stations in London (England) and at Munich airport (Germany) and depot-based facilities supporting the fuel cell bus fleets in Tochigi (Japan), Faridabad (India), Vancouver (Canada) and California (USA) (H2 Stations.org, 2006).

In the second category, a hydrogen carrier fuel is produced at a central location and distributed to fuel stations where it is processed to produce hydrogen on site. (A hydrogen carrier fuel is defined here as any fuel which can be used to generate hydrogen on demand.) In the UK, natural gas could be used to generate hydrogen on demand using small-scale reformers located at fuel stations. This would use the extensive natural gas grid that already covers a large proportion of the country. Similarly, electricity from the national grid could be used to electrolyse water to produce hydrogen where and when required. Many analysts have proposed this option as the most cost-effective method of hydrogen fuel infrastructure development, as it makes use of the existing fuel infrastructure to maximum effect (Hart et al., 2000).

Thirdly, a hydrogen-rich fuel can be processed on-board the vehicle (via catalytic reforming), thereby generating hydrogen gas on demand. This approach avoids the problems of hydrogen storage already discussed and reduces the need to build new fuel infrastructure, which would be required

Figure 2.20 Interior views of the hydrogen filling station in Reykjavik: on the left the electrolyser and compressor modules and on the right the main storage tanks and associated pipework

for a nationwide hydrogen gas network. In principle, any hydrogen carrier can be used for this option, although simpler hydrocarbons are easier to reform. As on-board reformers need to have fast response times, fuels that can be processed at low temperatures are preferred. Of the liquid fuels, methanol is unique in that it can be reformed into hydrogen at around 260 °C, as compared to 600–900 °C for other fuels such as petrol, ethanol, natural gas and propane. Therefore methanol is considered to be the prime candidate for on-board fuel storage and has been successfully demonstrated in test vehicles, including the Necar 3 and Necar 5 FCVs. However, many manufacturers are attempting to develop petrol reformers, driven by the possibility that FCVs may be able to utilise existing fuel infrastructure.

Environmental impact

Precise vehicle performance data is difficult to source owing to the current commercial sensitivity of fuel cell vehicle development. However, what data is available suggests that fuel economy is significantly improved for FCVs as compared to conventional vehicles owing to the high efficiency of the fuel cell drive-train in comparison with the ICE. Figure 2.21 shows estimates of energy use for a small FCV as compared to a petrol car. These figures are based on EU test and modelling data (Concawe, 2005).

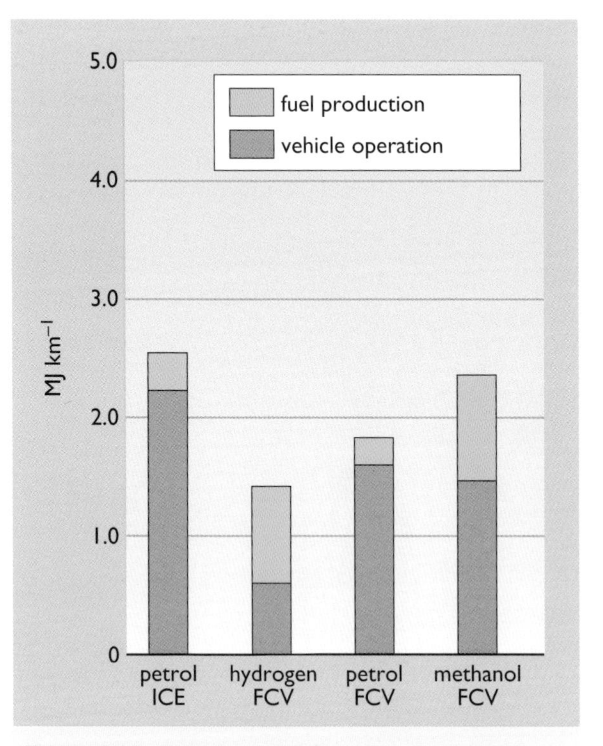

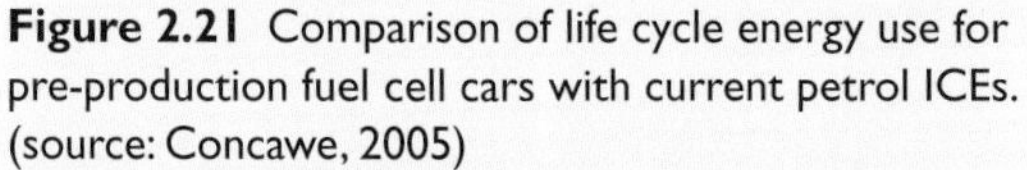

Figure 2.21 Comparison of life cycle energy use for pre-production fuel cell cars with current petrol ICEs. (source: Concawe, 2005)
Key: hydrogen FCV – hydrogen reformed at source, using natural gas feedstock; petrol FCV – hydrogen produced using on-board petrol reformer; methanol FCV – methanol produced at remote site using natural gas feedstock, hydrogen produced using on-board metal reformer.

If vehicle energy use and fuel production emissions data are combined, *life cycle* greenhouse gas emissions are predicted to be lower for FCVs than for their petrol equivalents, due to improved efficiency of the vehicle and fuel processing. However, the reduction is difficult to quantify, depending as it does on the method of fuel production. If natural gas is reformed on site at fuel stations, then modelling by Concawe suggests that greenhouse gases will be reduced by almost 60% on a life cycle basis for light-duty vehicles. Similar reductions are expected for a fuel cell bus. In principle, the use of renewable hydrogen would eliminate the emission of greenhouse gases altogether. However, this is likely to be an expensive option in the short term.

Estimates for *regulated* emissions suggest even greater life cycle emission reductions for methanol- and petrol-fuelled FCVs and FCVs using compressed hydrogen produced from on-site reforming of natural gas.

Safety concerns could act as a barrier to hydrogen fuel cell vehicle introduction (Boyle et al., 2003). However, there is a growing body of evidence to support the view that the use of hydrogen is no more dangerous than the use of petrol or other flammable fuels. In hydrogen's favour, the gas is non-toxic. A hydrogen fire produces no poisonous fumes and has a lower flame temperature than petrol-fuelled fires. Owing to hydrogen's low density, escaping gas rises away from a spill site, unlike petrol vapour (and LPG), which remains in the spill area so prolonging the fire's duration. With a high diffusion coefficient, hydrogen mixes in air faster than petrol vapour or natural gas, which is advantageous in the open (but could represent a potential disadvantage in a poorly ventilated enclosed space). Extensive destructive testing of pressurised hydrogen cylinders failed to produce any consequences as bad as those from comparable assaults on ordinary gasoline tanks (Williams, 1997).

Taken overall, while the risks of using hydrogen may not be greater than using conventional fuels, its use as a vehicle fuel requires *different* handling procedures. For example, hydrogen is colourless and odourless, which makes human detection difficult. Also, the gas burns in air at concentrations of 4–75% by volume (which is a larger range than for other fuels) and the minimum ignition energy required for a *stoichiometric* hydrogen/oxygen mixture is only 20 μJ, one order of magnitude less than for natural gas and petrol vapour (Hart et al., 2000). The use of hydrogen also causes some metals to become embrittled, and further research into the best types of stainless steel to replace these is ongoing.

Hydrogen – where next?

For hydrogen to become a significant transport fuel, there are many barriers that need to be overcome. Some of these barriers are summarised by Romm (2006) and relate to those already observed for other alternatively fuelled cars. They are issues directly linked to:

- high cost of vehicles
- limited vehicle (driving) range
- safety and liability
- high fuel cost
- limited fuel stations
- improvements in the incumbent technology.

These factors tend to limit the success of the new technology, but there are also broader issues at stake which were summarised by the National Academy of Sciences (2004) as presenting major challenges. These challenges have been rephrased here as questions, which will most likely remain at least partially unanswered for some time to come.

1 How long will the transition to hydrogen take?

2 What is the best way of converting our existing energy infrastructure to hydrogen?

3 Where will the hydrogen come from, and how can it be assured that it is CO_2 neutral (in the overall life cycle)?

Concerning issues 2 and 3, the sequestration of carbon is also a major challenge, especially if (in the case of the USA) coal is used a basis for the energy carrier.

Other issues highlighted include safety. The report concludes:

> Safety will be a major issue from the standpoint of commercialization of hydrogen-powered vehicles. Much evidence suggests that hydrogen can be manufactured and used in professionally managed systems with acceptable safety, but experts differ markedly in their views of the safety of hydrogen in a consumer-centered transportation system. A particular salient and underexplored issue is that of leakage in enclosed structures, such as garages in homes and commercial establishments. Hydrogen safety, from both a technological and societal perspective, will be one of the major hurdles that must be overcome in order to achieve the hydrogen economy.
>
> National Academy of Sciences, 2004

The report emphasises that within the transportation sector dramatic progress is required in the development of fuel cells, storage devices and distribution systems and that this progress is likely to take decades to achieve.

Although this chapter, and especially this last section has emphasised the importance of the role of technology in reducing our energy dependence, it is clear that behavioural changes will also become increasingly more important if we are to move away from oil-based mobility.

References

Alternative Fuels Data Center (AFDC) (2003) http://www.afdc.doe.gov/pdfs/fueltable.pdf [Accessed 18 August 2006].

American Coalition for Ethanol (ACE) (2006) http://www.ethanol.org.

Association Européenne Des Gaz De Pétrol Liquéfiés (AEGPL) (2006) http://www.aegpl.com/ [Accessed 27 July 2006].

Automotive Intelligence Data Ltd (AID) (2005) *Schmidt's Diesel Car Prospects to 2010*, Peter Schmidt, AID Ltd, Warwick CV34 4JP, UK, July 2005, (308 pages).

Avere (2006) European Association for Battery, Hybrid and Fuel Cell Vehicles, http://www.avere.org/ [Accessed 27 July 2006].

Bérnard, P. and Chahine, R. (2001) 'Modelling of adsorption storage of hydrogen on activated carbons', *International Journal of Hydrogen Energy*. vol. 26, pp. 849–55, Elsevier Science.

Boyle, G., Everett, B. and Ramage, J. (eds) (2003), *Energy Systems and Sustainability, Power for a Sustainable Future*, Oxford, Oxford University Press/Milton Keynes, The Open University.

Concawe (2004) *Well-to-Wheels Analysis of Future Automotive Fuels and Powertrains in the European Context.* Report by Concawe, Eurcar and the EU Joint Research Centre, Version 1.

Concawe (2005) *Well-to-Wheels Analysis of Future Automotive Fuels and Powertrains in the European Context.* Report by Concawe, Eurcar and the EU Joint Research Centre, Version 2.

Costlow, T. 'Power to the hybrids', *Automotive Engineering International*, June 2006, vol. 114, no.6, pp. 37–40.

Cuddy, M.R. and Wipke, K.B. (1997) *Analysis of the Fuel Economy Benefit of Drivetrain Hybridization*, NREL, Colorado, USA, http://www.nrel.gov/docs/legosti/fy97/22309.pdf [Accessed 14 July 2006].

CUTE report (2004) *Hydrogen Supply Infrastructure and Fuel Cell Bus Technology*, EvoBus, Ulm, Germany (46 pages).

Davenport, R.E. and Apanel, G (2004) *Issues Facing Global Methanol Industry*, Chemical Week conference presentation, Houston, Texas, October 25, 2004.

Davis, S.C. and Diegel, S.W. (2005) *Transportation Energy Data Book* (25th edn), Oak Ridge National Laboratory, Tennessee, USA.

Department for Transport (DfT) (2005) *Transport Statistics Great Britain*, DfT, The Stationery Office, London.

Department for Environment, Food and Rural Affairs, London (DEFRA) (2002), 'Air pollution – what it means for your health', http://www.defra.gov.uk/environment/airquality/airpoll/index.htm [Accessed 11 July 2006].

Department of Trade and Industry (DTI) (2000) *The Report of the Alternative Fuels Group of the Cleaner Vehicle Task Force Report*, Automotive Directorate, London, The Stationery Office.

DieselNet (2006) *Emissions Standards* (international) http://www.dieselnet.com/standards [Accessed 14 July 2006].

EAST (2005) 'Plug-in Prius', *Environmental and Sustainable Technology*, November.

Ecolane (2005) *Green Car Buyer's Guide*, Ecolane Limited. Available from www.ecolane.co.uk/ebook 2005.

Ecolane (2006) *Life Cycle Assessment of Vehicle Fuels and Technologies – Final Report*, March, pp. 13–15. Available from http://www.lowcvp.org.uk/uploaded/documents/Camden LCA Report FINAL 2005-06.pdf [Accessed 27 July 2006].

Energy Saving Trust (EST) (2001) PowerShift website, http://www.est-powershift.org.uk/ [no longer available].

European Automobile Manufacturers Association (ACEA) (2002) *Monitoring of ACEA's Commitment on CO_2 Emission Reductions from Passenger Cars* available at: http://www.acea.be [Accessed 27 July 2006].

European Biodiesel Board (EBB) (2006) http://www.ebb-eu org [Accessed 27 July 2006].

European Commission (EC) (2005) *Implementing the Community Strategy to Reduce CO_2 Emissions from Cars: Fifth Annual Communication on the Effectiveness of the Strategy*, COM(2005) 269 final, (see esp. Table 3, p. 11, 'Trends in new cars registered'), Brussels, Commission of the European Communities.

European Natural Gas Vehicle Association (ENGVA) (2006) http://engva.org [Accessed 27 July 2006].

Fraidl, G.K., Kapus, P.E., Piock, W.F. and Denger, D. (2000) 'Gasoline engine concepts related to specific vehicle classes', paper C588/009/2000, pp. 211–27, in IMECHE Conference Transactions, International Conference on 21st Century Emissions Control, © 2000, London, UK.

General Motors Corporation (GMC) (2003) *GM Hy-wire: Major Step Forward in Reinventing Automobile World's First Driveable Fuel Cell and by-wire Vehicle*, http://www.gm.com/company/gmability/adv_tech/400_fcv/hy-wire_overview_050103.html [Accessed 27 July 2006].

General Motors Corporation (GMC) (2005) http://www.cadillac.com, see sub page specifications, powertrain and options. http://www.cadillac.com/cadillacjsp/model/po_specification.jsp?model=cts&year=2006§ion=Powertrain [Accessed 27 July 2006].

Haraldsson, K., Folkesson, A. and Alvors, P. (2005) 'Fuel cell buses in the Stockholm CUTE project – first experiences from the climate perspective', *Journal of Power Sources*, vol. 145, pp. 620–31.

Haraldsson, K., Folkesson, A., Saxe, M. and Alvors, P. (2006) 'A first report on the attitude towards hydrogen fuel cell buses in Stockholm', *International Journal of Hydrogen Energy*, vol. 31, pp. 317–25.

Hart, D. and Bauen, A. (1998) *Fuel Cells: Clean Power, Clean Transport, Clean Future*, Financial Times Report, Financial Times Energy.

Hart, D., Fouguet, R., Bauen, A., Leach, M., Pearson, P. and Anderson, D. (2000) *Hydrogen Supply for SPFC Vehicles*, ETSU, UK (ETSU F/02/00176/REP).

HyWeb (2001) HyWeb website, http://www.HyWeb.de/ [Accessed 18 August 2006].

H2 Stations.org (2006) 'Hydrogen filling stations worldwide', http://www.h2stations.org/ [Accessed 20 June 2006], maintained by Ludwig-Bolkow-Systemtechnik GmbH, Ottobrunn, Germany, www.lbst.de

International Energy Agency (IEA) (1999) *Implementation Barriers of Alternative Transport Fuels*, IEA/AFIS, Innas BV, February 1999.

International Energy Agency (IEA) (2005) *Key World Energy Statistics*, Paris, http://www.iea.org/dbtw-wpd/Textbase/nppdf/free/2005/key2005.pdf [Accessed 14 July 2006].

Johansson, T.B., Kelly, H., Reddy, A.K.N. and Williams, R.H. (eds) (1994) *Renewable Energy: Sources for Fuels and Electricity*, London, Earthscan, Island Press.

Kitman, J.L. (2000) 'The secret history of lead – a special report', *The Nation*, 20 March.

Lenton, T.M., Loutre, M.F., Williamson, M.S., Warren, R., Goodess, C.M., Swann, M., Cameron, D.R., Hankin, R., Marsh, R., and Shepherd, J.G. (2006) *Climate Change on a Millennial Timescale – phase 1: coarse scale climate prediction*, http://www.environment-agency.gov.uk/commondata/acrobat/tyndall_1298654.pdf [Accessed 27 July 2006].

Liquefied Petroleum Gas Association (LPGA) (2006) http://www.lpga.co.uk/ [Accessed 27 July 2006].

LowCVP (2006) 'European car CO_2 emissions fall in 2005, but 2008 target out of reach says T & E', LowCVP online News, April 2006. http://www.lowcvp.org.uk [Accessed 27 July 2006].

MacLean, H.L. and Lave, L.B. (2003) 'Evaluating automobile fuel/propulsion systems technologies'. *Progress in Energy and Combustion Science*, vol. 29, no. 1, pp. 1–69.

McKay, A. (2006) 'G-Wiz – our urban friend's electric', *Scotland on Sunday*, 30 April 2006, http://scotlandonsunday.scotsman.com/index.cfm?id=645212006 [Accessed 27 July 2006].

Monaghan (1998) 'Particulates and the diesel – the scale of the problem' (paper S491/001/98), *Diesel Engines – Particulate Control*, IMechE Seminar Publication, p. 16.

Mildenberger, U. and Khare, A. (2000) 'Planning for an environment-friendly car', *Technovation*, vol. 20, no. 4, pp. 205–14.

National Academy of Sciences (2004) *The Hydrogen Economy – Opportunities, Costs, Barriers and R&D Needs*. The National Academies Press, Washington, DC.

Natural Gas Vehicle Association (NGVA) (2006), http://www.ngva.co.uk [Accessed 27 July 2006].

Pearson, J.K. (2001) *Improving Air Quality: Progress and Challenges for the Automotive Industry*, Warrendale, PA, USA, Society of Automotive Engineers.

PriceWaterhouseCoopers, News Release, *Diesel Car Sales Set to Overtake Petrol in Europe,* 30 January 2006 (6 pages).

Renewable Fuels Association (RFA) *Ethanol Production Statistics* (global) RFA, Washington, DC, http://www.ethanolrfa.org/industry/statistics/ [Accessed 14 July 2006].

Romm, J. (2006) 'The car and fuel of the future', *Energy Policy*, in press, pp. 1–6.

Schulte, I., Hart, D. and van der Vorst, R. (2004) 'Issues affecting the acceptance of hydrogen fuel', *International Journal of Hydrogen Energy*, vol. 29, pp. 677–85.

Serfas, J. A., Nahmias, D. and Appleby A. J. (1991) 'A practical hydrogen development strategy', *International Journal of Hydrogen Energy*, vol. 16, no. 8, pp. 551–6.

Society of Motor Manufacturers and Traders Limited (UK) (SMMT) (2006) *UK New Car Registrations by CO_2 Performance, a report based on the 2005 Market*, April 2006 (61 pages), London, http://lib.smmt.co.uk/articles/sharedfolder/Publications/ACF22CC.pdf [Accessed 20 June, 2006].

Stone, R. (1999), *Introduction to Internal Combustion Engines*, London, Macmillan.

Teufel, D., Bauer, P., Lippolt, R. and Schmitt, K. (1993) *OeKO-Bilanzen von Fahrzeugen*, 2. erweiterte Auflage, Heidelberg: Umwelt-u. Prognose-Institut Heidelberg e.V.

Toyota *Worldwide Prius Sales Top 500 000 Mark*, 7 June 2006, http://www.toyota.co.jp/en/news/06/0607.html (data source for Toyota Prius Sales Globally) [Accessed 13 June 2006].

US Department of Energy (2006a) *FreedomCAR and Vehicle Technologies Programme*, http://www1.eere.energy.gov/vehiclesandfuels/, [Accessed 18 August 2006].

US Department of Energy (2006b) *FreedomCAR and Vehicle Technologies Multi-Year Programme Plan*, Chapter 3, 'Goals', http://www.eere.energy.gov/vehiclesandfuels/pdfs/mypp/3_goals.pdf, [Accessed 18 August 2006].

US Department of Energy (2006c) *Partnership Plan – Freedom Car and Fuel Partnership*, see especially Figure 3 (p. 10), http://www.eere.energy.gov/vehiclesandfuels/pdfs/program/fc_fuel_partnership_plan.pdf [Accessed 27 July 2006].

Vehicle Certification Agency (VCA) (2006) http://www.vca.gov.uk/ [Accessed 27 July 2006].

Wainwright, M. (2002) 'Chicken Fat to Power Lorries', *The Guardian*, 29 October.

Wells, P. (2006) 'Does HCCI have a future?' *Automotive Analyst*, vol. 128, February 2006.

Williams, B.D. (1997) 'Hypercars: speeding the transition to solar hydrogen', *Renewable Energy*, vol. 10, no. 2/3, pp. 471–9, UK, Pergamon. Adapted from Lovins et al. (1996) *Hypercars: Materials, Manufacturing and Policy Implications*, Colorado, Rocky Mountain Institute.

Zakian, M. (2006) 'Gentlemen, charge your engines', *The Guardian*, 15 June.

Chapter 3

Mobility management in organisations

by Stephen Potter and Marcus Enoch

3.1 Transport impacts and institutions

In Chapter 1, it was concluded that the only potentially viable approach to dealing with the energy and environmental impacts of transport is to combine technical improvements in fuel efficiency and lower carbon fuels with changes in people's travel behaviour. This follows the concept of 'intelligent consumption' – the idea that the benefits achieved by travel can be obtained at a lower energy and resource cost. Chapter 2 looked at technical methods of reducing emissions of pollutants, improving fuel efficiency and introducing lower-carbon fuels. In this chapter and Chapter 4 we will look at the changes that can be made in our patterns of travel to reduce energy use and environmental impacts. This is not to say that technology has no role in adapting travel behaviour. Indeed, the development of certain key technologies is crucial to the success of such approaches. Adapting travel behaviour frequently requires technologies to support it, just as improving fuel efficiency and moving to lower-carbon fuels may require some behavioural change on the part of vehicle users.

BOX 3.1 Transport terminology

The concept of managing the demand for transport is known by several names. In the USA the term **transportation demand management** (frequently abbreviated to TDM) is used. In Australia and the UK the variant **transport or travel demand management** is used. Among EU policymakers, **mobility** management is the most common term, and is now being increasingly used in the UK. This is the term adopted here as it covers the full range of transport planning mechanisms.

Although transport policy is often viewed as something done by Government that affects individuals, it is becoming increasingly important for institutions such as employers, shopping centre managers and big service providers to have a role in transport policy. This is not an easy policy area, as many people are reluctant to accept what they see as infringements upon travelling in the way they desire, and organisations do not see it as part of their business to 'interfere' with the travel behaviour of their staff. Despite this, the role of employers and other institutions in supporting more sustainable transport policies is a new and important field (see Figure 3.1). An example of an institutional response was given in Chapter 2 (Box 2.9), which described the introduction by Oxfordshire Mental Healthcare NHS Trust of cleaner liquefied petroleum gas (LPG) vehicles. However, rather than looking at cleaner vehicle technologies, this chapter and the next will focus on the potential for institutions to manage and reduce the volume of travel generated by their staff and customers.

Figure 3.1 Organisations located in profitable city-edge locations have major transport impacts. Increasingly they are being asked to take some responsibility for the traffic and environmental effects generated

3.2 The 'bed of nails' of mobility management

The transport policy challenge

Back in 1981 the development of an 'integrated transport policy' was the subject of an episode of the BBC political comedy *Yes, Minister*, entitled 'The bed of nails'. The plot was that transport involved so many irreconcilable desires, interests and approaches that the only politically viable transport policy was not to have a policy at all, least of all an integrated transport policy. More than 25 years later that has proved to be an astute observation. At the heart of the transport crisis is a widespread paradoxical reaction. While, on the one hand, people accept that our high and growing level of road transport dependence is costly, unpleasant, unsustainable and generally undesirable, they resist policies that actually affect themselves, their town or their organisation.

Behind this is a tension between collective effects and individual benefit. While increased reliance on cars has caused all sorts of problems for society as a whole, on an individual level the use of the car confers substantial benefits. Convenience, comfort, flexibility, personal space, and low perceived cost are often cited as reasons for the dominance of travelling by car, but other deeper reasons have been suggested (see Figure 3.2).

> The sensual, erotic, or irrational well springs of the auto mobility cannot be ignored. The pleasure, as well as the convenience that auto driving provides is a boon to many people. However, what is needed is a transport system that allows people to find pleasure in many ways of travel. New policies must be as non-punitive as possible in discouraging auto use, and must develop seductive, as well as affordable and efficient alternatives to the auto.
>
> Freund and Martin, 1993

Figure 3.2 Car advertisements often play on emotion and status.

Furthermore, for many people there is no real viable alternative to using their car. We have gradually adjusted our lifestyles over the years to depend on the personal mobility that a car provides. The car has generated journeys and a lifestyle for which public transport is not as convenient (if even possible), while walking and cycling are not feasible because distances for many journeys are now too great.

The crux of the problem is that the benefits of car use are very evident to individuals, whereas the problems are more diffuse and affect the population as a whole, with some impinging on future rather than current generations. This means that any individual action to reduce car use produces little or no obvious personal benefit. For example, one parent letting his or her children walk to school will not improve their safety, so long as no similar action is taken by other parents. Furthermore, the cut in pollution will seem negligible and any benefits may not be felt for many years. Any global benefits may not even affect the original country. This unequal conflict between choosing immediate and tangible personal benefit over a delayed, dispersed and far less visible cost to society is behind many of the difficulties faced in addressing the transport crisis. It is thus not surprising that transport policy is, politically, a bed of nails!

Targeting the 'easy wins'

Policies to manage travel, although necessary for a variety of reasons, are thus far from being politically popular. Policy makers therefore feel they need to go for the 'easy wins' and target those people most likely and able to alter their car use and to produce the most obvious benefits. An example may be park-and-ride sites at the edge of historic cities such as York, Oxford and Chester. There is an acceptance that traffic adversely affects the quality of life in the historic centre of these cities and that the narrow city-centre roads cannot be widened. The park-and-ride schemes target shoppers and commuters, for whom the change in behaviour is relatively easy.

The identification of appropriate groups of people or types of journey can be done in a number of ways. One method is to use socio-economic data to identify the type of car user most likely to walk, cycle or use public transport, in a similar way to supermarkets' use of market research companies to match people to products. This technique is now slowly being adopted by the bus industry in Britain and in various towns and cities across the world (including Perth, Western Australia and Leeds, West Yorkshire). Termed **travel blending**®, it involves identifying and 'educating' those most capable of switching from the car to other modes of transport. Box 3.2 provides more information about this approach.

BOX 3.2 Travel blending

Travel blending – What is it?

Travel blending is the terminology used to describe a way for individuals to reduce the use of the car, which involves:

- thinking about activities and travel in advance (i.e. in what order can I do things, where should they be done, who should do them?)

and then:

- blending modes (i.e. sometimes car, sometimes walk, sometimes public transport, etc);
- blending activities (i.e. doing as many things as possible in the same place or on the same journey); and
- blending over time (i.e. making small sustainable changes over time – once a week or once a fortnight).

People and households who take part in travel blending choose to change their behaviour by:

- observing their own current travel patterns – measuring the way they and their households use the car for one week;
- receiving detailed suggestions customised to those travel patterns;
- setting their own targets;
- spending some weeks trying to reduce the use of the car;
- observing the changes they have achieved;
- being given a simple, ongoing system of monitoring and motivation.

Travel blending – Where has it happened?

UK: Bristol, Darlington, Frome, Hastings, Leeds, London, Nottingham, Petersborough

USA: New Jersey

Australia: Adelaide, Sydney

Chile: Santiago.

Reductions in car driver trips for those participating ranged from 6–23%. The results for a pilot scheme in Nottingham are shown in Table 3.1. Participants kept a travel diary of trips made before and after the travel blending exercise. The column for 'whole population' includes those who refused to take part or dropped out of the programme. Even allowing for such non-participation, useful cuts in car use were achieved.

Table 3.1 Results of a pilot scheme to reduce car driver trips in Nottingham

	Diary 1	Diary 2	Change of participants (%)	Change in whole population (%)
Car driver trips	19.1 trips	17.7 trips	–7.6%	–3.3%
Car driver miles	147.3 miles	126.5 miles	–14.2%	–6.2%
Total hours in the car (all resp.)	7.5 hours	6.6 hours	–11.8%	–4.8%

Steer Davies Gleave, n.d.

Another approach is to identify not the people, but the reasons for the journeys being made. This involves considering why people make particular journeys and then trying to devise ways of reducing certain types of trip. For example, it may be possible to reduce the number of shopping trips by car through introducing teleshopping, internet shopping or home delivery schemes. The number of school journeys made by car might be cut by building safe cycle routes, operating school buses or organising supervised walking groups (called 'walking buses').

In theory, some of the easiest journeys to deal with are those that people make every day, e.g. commuting trips to and from work. These usually have a fixed start and end point (home and work), and are generally made at similar times each day. Commuting and business trips account for 18% of the total number of journeys made in the UK. Of these, 71% are made by car (DfT, 2005). Any change in such behaviour therefore would make a sizeable impact on the transport problem.

Targeting commuter trips will frequently require the involvement of employers. So what role can employers play in changing the travel behaviour of their staff to reduce environmental impacts? From the late 1970s in the USA, transport demand management by employers became part of a range of initiatives to improve air quality. It was implemented by regulations requiring employers to cut car commuting (particularly single-occupant car commuting) to their sites. This approach is now being developed in Europe, with a mixture of regulations, tax incentives and voluntary agreements seeking to stimulate employers to introduce measures to help their staff commute in a more environmentally friendly way.

BOX 3.3 More on transport terminology

It has already been noted in Box 3.1 that the concept of managing the demand for transport is known by several names, and in this text the term 'mobility management' is mainly used. Within the general policy approach of mobility management, there are a number of measures, including employer-led initiatives. Such initiatives have also been referred to by a variety of terms, including **commuter plan**, **green commuter plan**, **mobility plan** and **green transport plan**. In the UK, the term 'travel plan' is now the most common term, and we shall use it in this chapter and in Chapter 4, although you should be aware of the varying terminology. Mobility management and travel plans are new areas and the terminology has yet to settle down.

3.3 What is a travel plan?

For many years some employers have helped staff travel to work in one way or another. Factories in remote locations might provide 'works buses' from nearby towns, and often special transport arrangements are made for late-night shift workers. However, today it is more likely to be providing a company car or large, free car parks. The travel plan concept – that employers and organisations should take a responsibility to manage how staff, customers and visitors travel to their site in order to address wider public policy reasons – is a recent development. However, its history can be traced back to the 1940s, when as part of the war effort to conserve fuel, car sharing and measures to reduce 'unnecessary' trips were promoted. Employers took part in this campaign and the first organised company car sharing scheme was introduced in the USA by Boeing in Seattle. A car sharing poster of the time declared 'When you ride ALONE, you ride with Hitler!".

These early 'travel plans' were, of course, nothing to do with environmental sustainability, but about contributing to the fight against aggressive totalitarian regimes. One might say that, today, we are in an analogous situation to that faced in the Second World War. The battle to achieve environmental sustainability is at least as serious as then, and we are only just beginning to appreciate the scale of disaster that defeat would bring. So the travel plan has emerged as one weapon in the campaign for everybody to contribute to the war for sustainability. The first environmental requirements for employers to manage the travel of their staff came as part of Californian air-quality legislation in the 1970s. The Dutch adopted travel plan type measures in the 1980s and local authorities in the UK (notably Nottingham City Council) began promoting 'green commuter plans' in the early 1990s. The promotion of travel plans became UK Government policy from 1997, with workplace and school travel plans featuring in the 1998 and 2004 transport policy White Papers (DETR, 1998 and DfT, 2004). Government guidance defines a travel plan as being:

> A general term for a package of measures tailored to [meet the] needs of individual sites and aimed at promoting greener, cleaner travel choices and reducing reliance on the car. It involves the development of a set of mechanisms, initiatives and targets that together can enable [an] organisation to reduce the impact of travel and transport on the environment, whilst also bringing a number of other benefits to [the] organisation as an employer and to staff.
>
> Energy Efficiency Best Practice Programme, 2001, Section 1.1

The key point about travel plans is that those organisations responsible for creating the need to travel, such as employers, service providers and shopping centre owners, are involved in helping to solve transport problems. The involvement of such institutional players is both a strength and a weakness of the travel plan approach. The main weakness is that the vast majority of employers and other institutions do not see solving transport problems as their responsibility. To date, rather than adopting an integrated management approach, employers have tended to treat transport matters as separate, self-contained issues, many of which are seen as largely outside an employer's control and thus not their responsibility. These issues have included:

- road congestion affecting delivery reliability and costs (as well as staff punctuality)

- congestion of on-site parking
- transport-related planning conditions required for site development
- changes to the tax treatment of transport benefits in the remuneration package (company cars, mileage allowances, etc.) and other transport-related human resource issues
- transport and company environmental policies (including environmental requirements of export markets and other supply chain pressures)
- transport effects upon brand image and public relations
- transport impacts upon company quality initiatives.

Figure 3.3 Traffic congestion affects delivery, reliability and staff punctuality

Institutional aspects of transport are now part of the UK Government's integrated transport policy. The rise in congestion and pollution is not bad just for the environment and society – it is bad for business as well. Congestion costs money and a variety of new measures are planned or under way that will have impacts upon employers, including:

- workplace parking charges (whereby some cities plan a levy on each parking space on an employer's site)
- congestion charging (where motorists are charged to enter a city centre, as in London (£8), Durham (£2), Oslo (£1) and Singapore), also known as **road user charging** or **area licences**
- changes in company car taxation to favour 'greener' vehicles
- changes in general vehicle taxation to favour fuel efficiency and cleaner fuels for both cars and trucks
- measures to increase the choice and opportunity for travel by 'greener' forms of transport
- tax concessions to employees for some employer-provided 'green' transport.

As in all good management practice, it is crucial not to treat these seemingly disparate issues in isolation: this leads to ad hoc 'fire fighting' that is costly and ineffective. The key to a cost-effective and successful approach is to recognise that organisations are dealing with a series of challenges and opportunities stemming from key changes in the transport environment. A strategic, integrated business approach is needed, and this is where travel plans come into their own.

Nevertheless, companies generally consider travel plans only when some other pressing reason forces them to examine how their staff get to work (Rye, 2002). There are notable exceptions – for example, The Body Shop actively wants its brand to be identified as being 'environmentally aware', and so it has adopted a travel plan for many of its sites in the UK. In some cases a travel plan can arise from a crisis of on-site parking congestion, or a perception that a company's transport problems are harming its business image. For example, it could be embarrassing to a university that teaches environmental management in its courses, if it could not show that it practises what it preaches by having an effective travel plan. However, in the majority of cases the most pressing reason arises when a company wishes to move into a particular area or expand its site, and the local authority forces it, through a condition of planning consent, to develop alternative ways in which employees or customers may travel. Such a planning condition is often called a **Section 106 agreement**, after the section of the 1990 *Town and Country Planning Act* that provides powers for councils to set such conditions. In Scotland this is called a **Section 54 agreement**, after the section in the comparable Act for Scotland.

For some public sector organisations, a travel plan is now required directly by Government or as a condition of finance. When applied well, travel plans can cut car use by worthwhile amounts. The best employer travel plans in the UK have secured a reduction in car use of between 10% and 20%, while some feel that up to 30% is a possibility (Cairns et al., 2004). In the USA, where mandatory travel plans have been in use, a 30% cut in car use has been achieved in several cases.

Incentive mechanisms for employers

In addition to providing information and guidance on why and how to introduce travel plans, there are effectively three other mechanisms to persuade companies to encourage their staff to commute in a 'greener' way. These are:

1 regulation

2 subsidies

3 the tax system.

Regulation

Although in the UK a travel plan may be required to obtain planning consent for a site development, in some countries a whole regulatory framework governs how companies deal with their employees' commuting. We have already noted that air-quality legislation in the USA included a mandatory requirement for larger employers to reduce driver-only car commuting to specified levels. This is now no longer required at the federal level (tax incentives are used instead), but several individual states have retained a regulatory requirement. In some places employers are required by law to subsidise their employees' public transport costs. For example, since 1983, in the Paris region of France, employers have been required to refund half the cost of the *Carte Orange* season ticket (see Figure 3.4) (Flowerdew, 1993). A similar scheme, the *Vale Transporte*, operates in Brazil.

In Italy, the government has begun pursuing a mandatory approach to travel plans. In 1998 the Environment Ministry mandated the Decree on Sustainable Mobility in Urban Areas. Organisations employing over 300 staff must designate a mobility manager to coordinate efforts to reduce employees' home–work trips through a site-specific 'mobility plan'. However, the impact of this measure is limited as no quantitative targets are set and there are no penalties for companies that do not comply (MOST, 2001).

Figure 3.4 The Paris Metro and (inset) the Carte Orange. Employers in Paris must subsidise the public transport tickets of their staff

Subsidies

An alternative to regulation is to use public subsidies as an inducement. In most cases subsidies are used to help organisations develop their travel plan programmes. For example, the Space Coast Area Transit Agency in Southern Florida, USA supports the Space Coast Commuter Assistance (SCCA) programme to help commuters use alternative transport to the private car. The agency assists businesses individually to develop programmes for reducing commuter trips, and makes no charge for its services (Litman, 2001). In Linz, Austria, the city council offers a free mobility consulting service to the 450 companies that have 50 employees or more (Schippani,

2002). In the UK, until 2006, the government funded a free travel plan advisory service for employers and schools, providing 5–10 days of expert advice to each site. This has now been replaced by a local system of School Travel Advisory and local authority Travel Plan Coordinators, who provide travel plan development advice. Direct cash subsidies are less common. One example is in Montreal, Canada, where cash subsidies have been offered to employers with more than 50 employees to develop travel plans (Coulliard, 2002). In Italy, the Environment Ministry set aside €15.5 million over three years to finance up to 50% of the design and implementation costs of companies' mobility plans (MOST, 2001).

Another form of subsidy is when public transport bodies and/or local authorities offer employers discounts for buying public transport passes. The Milton Keynes Transport Partnership runs one such subsidy scheme whereby employers with a travel plan can provides their staff with a free bus pass for one month, followed by a half-price pass for the second month and a 25% saving on an annual bus pass (see Figure 3.5). Travel West Midlands and the local Passenger Transport Executive, Centro, offer a 50% reduction on annual bus and rail travelcards to employers taking part in Birmingham's *Company Travelwise* scheme, but this is only if the employers withdraw free staff parking (!) There are several other schemes like this in the UK and in other countries.

Figure 3.5 Employers with travel plan in Milton Keynes can provide their staff with discounted bus tickets

Employers may also negotiate ad hoc discounts with bus and rail operators, but these are a business arrangement rather than a direct subsidy. Typically these deals may cut public transport costs by about 10%, but in some situations more substantial discounts can be organised. For example, Agilent Technologies on the edge of Edinburgh negotiated a 45% discount on rail weekly season tickets. However this was possible only because travel to their city-edge site was in the opposite direction to peak flows on the line into Edinburgh. It was therefore in the rail company's interest to fill empty seats at a discounted rate. City-centre employers have not been able to negotiate such a good deal, but the Edinburgh Chamber of Commerce has negotiated a 10% discount on annual bus season tickets for some employers in the city.

Tax System

In the UK, Ireland and the USA, the tax system views commuting as a private activity, and hence any employer support to commuting is liable to be taxed as 'income in kind'. There are some exceptions to this, the main one being the provision of parking for employees' cars, but until recently any support provided for public transport was liable to be taxed. Legislation in the USA provides tax exemption for employers to subsidise the public transport fares or **vanpool** costs of their staff (a vanpool is a company-owned minibus, driven by an employee who picks up others in their 'pool' on the way into work) by $100 a month (plus $155 for vanpool parking). If the employer does not subsidise public transport fares, individuals can buy tickets free of tax up to a specified allowance (IBI Group, 1999). Since 1999 the UK has introduced a series of tax concessions to support travel plans. A number of important travel plan measures have now been removed from the tax net, including private works buses, subsidies to improve the quality and coverage of bus services to an employer's site, and the provision of bicycles (Potter et al., 2003). Bus fares can also be subsidised on routes to an employer's site, but subsidies for train, metro or tram fares remain taxable.

Ireland has adopted a simpler and more comprehensive approach. In 2000 a tax reform was introduced whereby company provision of monthly or annual public transport season tickets became tax free.

In most mainland European countries the tax treatment of commuting is different. Commuting is a tax-deductible expense and any employer support for commuting is already tax free. For example, in Germany, up to 2001, commuters could deduct a generous kilometre rate for car commuting, whereas public transport commuters claimed the actual fares paid. Motorists felt they 'made money' on the tax relief and this tax system was widely viewed as encouraging car commuting. Consequently, in 2001 the tax rules were changed to provide the same kilometre rate for car and public transport. In 2004 the Netherlands similarly introduced reforms to provide the same tax relief rate for car and public transport commuting, together with some additional allowances to favour public transport and cycling commuting.

Overall, as detailed in Potter et al. (2006), these tax reforms in the UK, Ireland, the USA, Germany and the Netherlands have provided additional tax relief to employees for 'greener' commuting and travel plan measures. These reforms produce tax benefits to staff, but not to their employer. Consequently, if the employer feels they have little to gain in providing tax-free travel benefits to their staff, and do not do so, then the employee tax concessions will count for nothing. This is a weak link, and it seems likely that future tax concessions may need to concentrate on employers to complement the existing tax concessions to their staff. One example of such an approach comes from the USA state of Oregon, where businesses can receive a 35% tax credit for their investments in trip-reduction activities, including teleworking equipment for their employees, vehicles for vanpooling and bus passes (Litman, 2001). In the UK this could take the form of higher tax relief on specified travel plan expenditure. There is also an issue of tax incentives for developers to provide more sustainable transport infrastructure – for example providing higher tax relief on cycle facilities, provision for buses, etc. and a cap on the amount of expenditure on car parks that can be set against corporate tax liability.

3.4 Transport impacts at hospitals

One employer that has come to view transport as very much its own problem is the National Health Service (NHS). As the NHS sees it, 'Effective transport management is essential to minimise the negative environmental impact of healthcare related transport. Ambulances, patients, visitors, staff, suppliers, contractors arriving and leaving from healthcare facilities and vehicles deliver[ing] community-base healthcare can all lead to congestion, pollution and increased numbers of road traffic accidents' (NHS Estates, 2006). Hospitals attract a great number of visitors and have a large number of employees. They also need to have good access for ambulances, to be accessible to people who have a variety of disabilities, and to receive a variety of deliveries. At the same time hospitals should provide a tranquil atmosphere, where people can convalesce. The locations of hospitals and other health institutions are often far from ideal in meeting all these needs. The NHS is the largest employer in Britain, with around a million staff. On top of that, it generates a million patient journeys each day. Hospitals are the largest generators of traffic outside peak hours and are estimated to account for up to 5% of all trips (DoT, 1996). With around 70% of trips to and from hospitals being made by car, the NHS contributed 2.1 billion car trips to Britain's roads in 2000 (Dublin Transportation Office et al., 2001).

Figure 3.6 Hospitals are the largest generators of traffic outside peak hours

The general transport impacts of hospitals may be large, but transport is also an important internal cost to hospitals and a significant factor in the drive for efficiency improvements in the NHS. The 2004 guidance on accessibility planning (DoH, 2004), noted that 20% of all people (and 31% of those without access to a car) and more than half of older people found it difficult to travel to a hospital. This report advocated effective hospital travel plans linked into the Local Transport Plans of their council.

Other travel-related problems include:

- patients missing appointments
- visitors who are already under stress because a relative or friend is suffering in hospital have added frustration and aggravation through inability to park or the cost of parking, especially for long-term hospital stays
- staff complaining that they cannot get to work and/or the cost of parking being too high
- local residents complaining that they cannot get to their own homes because hospital staff and visitors park in front
- local authorities complaining because roads become congested with parked cars, causing hindrance and inconvenience
- potential for bad congestion and poor traffic flows hindering ambulances getting patients to the hospital or Accident and Emergency (A&E) facility
- likelihood of increased road traffic accidents.

In addition, providing new parking spaces is very expensive. Each car parking space costs between £300 (surface car parks) and £3000 (multi-storey car parks) per annum – money that might be better spent on improving the NHS.

The transport circumstances of hospitals were explicitly recognised by the Government in the 1998 White Paper *New Deal for Transport: Better for Everyone*. This stated that:

> [the Government is] particularly keen that hospitals are seen to be taking the lead in changing travel habits. By the very nature of their work, hospitals should be sending the right message to their communities on acting responsibly on health issues. We would like to see all hospitals producing green transport plans.
>
> DETR, 1998, p. 141

Hospitals are at or near the top of the list of institutions required to contribute towards solving the transport crisis. Within the NHS itself, the desire to reduce the transport problems caused by healthcare sites is reiterated in a number of policy documents. For example, the *New Environmental Strategy for the National Health Service* (NHS Estates, 2002) identified transport as one of five key areas where progress needs to be made; the others are procurement, energy, waste and water. Specifically, the document stated that NHS sites should have developed a 'healthy travel plan' by October 2002. This should be compatible with the local authority transport strategy and identify the potential for reducing journeys and using smaller-engined, low-sulphur or LPG-fuelled vehicles. The benefits would be increased fuel economy and lower tax, less pollution, better financial returns and less stress from traffic jams. It is interesting to note that these benefits arise from a mixture of technical measures and behavioural change measures. Travel

to 54% in 2001. In particular, bus use had more than doubled from its 1995 share of 8% of trips.]

...

Costs and benefits

The annual cost of funding the travel plan is approximately £150 000. In 2001 this was comprised of:

- Car sharing — £200
- Bus measures — £59 500 (+£17 500 for national insurance)
- Publicity and promotion — £3 000
- Cycling measures — £15 000
- Staff time in managing the plan — £16 500
- Bike/motorcycle interest free loan — £6 480
- The remaining £31 820 is spent on maintenance, security, lighting, landscaping, pavements etc.

There have been some initial costs for setting up aspects of the travel plan in 1997–98 and in total it is calculated that this was £127 000.

The trust has also calculated that the upkeep and day to day operation of the site's car parking facilities, including demand management measures, costs the trust £445 000 per annum (i.e. £210 per space) at 2000 prices.

Annual running cost per member of staff (calculated as £150 000/4193 full time equivalent) is just under £36.

Support for bus and rail use

There is no local railway station. The main Plymouth station is located 5 miles away in the city centre.

Prior to the development of the travel plan in 1997–98 there were 22 bus services serving the site at peak hours. By 2001 this figure had risen to 44 buses. Derriford Hospital is consequently well served by public transport. The bus operators have restructured their services so that 80% of the existing routes serving the northern part of Plymouth provide direct and frequent access to the hospital. In collaboration with the city council and the bus companies, the trust has joint funded and produced a Travel to Derriford leaflet with bus timetables.

In 1997 there were two bus stops on site. This has risen to five with three bus shelters and set up and set down points. The hospital's bus lay-by has been trebled in size in order to cope with the higher volumes of bus traffic [and the] trust has ... agreed to the creation of a purpose designed bus station on the hospital site. This is being funded through revenue from Plymouth City Council's local transport plan.

There is an array of discount subsidised bus passes available. The original was the Derriford Travel Pass available to staff handing back their car parking permits. This involved a half price ticket, which was 40% subsidised by the trust and 10% by the public transport operator. From April 2000 the trust has offered a four-month trial free bus pass to staff for handing back a car-parking permit. At the end of the four months staff continuing to use the bus can get a 65% reduction on a bus pass for 12 months of which 55% is trust subsidised and 10% from the public transport operator. Further bus passes have a 50%

(a) (b) (c)

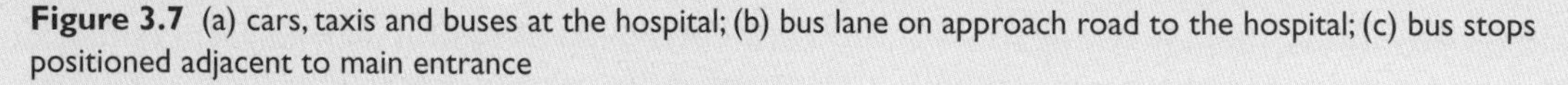

Figure 3.7 (a) cars, taxis and buses at the hospital; (b) bus lane on approach road to the hospital; (c) bus stops positioned adjacent to main entrance

discount. By 2000, 443 discounted tickets paid for by the trust amounted to £80 130.

For other staff, there is a Green Zone Bus Pass, introduced from April 2000, comprising of five zones. This discounted ticket has been negotiated with other local employers and the city council. The Green Zone Bus Pass gives a 25% reduction on the standard single bus journey ticket. For journeys within 5 miles a monthly ticket costs £29.25, for 5–10 miles £36.00, for 10–15 miles £42.50 a month, 15–20 miles £48.00 and 20–25 miles £53.00. The tickets are valid for bus services provided by both main operators in the city, Plymouth City Council, and First Western National. Other ticket offers include 10 journeys for the price of 12 and again these are valid with both the main bus service providers.

The trust has funded discounts on two routes to the hospital operating through areas of poor health. In addition, it has encouraged visiting between 6 and 8 pm through cheaper parking rates. The latter has resulted in a 25% increase in evening visiting since 1999.

Support for cycling

Access to the site is reasonably good by bicycle. There have been both off and on-site improvements for cyclists since 1997. Off site measures have been developed independently by Plymouth City Council. On site facilities include shortened road humps so that cyclists can avoid these. There were existing showers and changing rooms which can be used by cyclists and extra lockers were introduced in 1997–98. In 1997 there were no bicycle parking facilities but by October 2001 there were 100 spaces.

From 1998 the trust has offered staff a £500 three-year interest-free loan for the purchase of a bicycle. Cycle training is offered to staff but there have been no demands for this. The trust has produced a one off newsletter in June 2000, Pedal Power.

Support for walking

Accessibility of the site by foot is described as 'medium'. The trust has pressed the city council to make improvements to off site pedestrian facilities. Regarding on site facilities, the trust has completed development of a pavement network. In 1998–99 five zebra crossings were installed (and one removed). The trust has also funded improved lighting.

As with cyclists, pedestrians have access to showers, changing facilities and lockers. There is a contracted security patrol service operating across the site.

Support for car sharing

The trust has operated a computerised matching service since 1997–98. In 2000 the names of 640 potential car sharers were contained on the computer database. Car sharers are exempt from car parking charges and have priority parking spaces closest to the hospital buildings. There are 130 car parking spaces reserved for car sharers ... Since April 2000 there is a guaranteed [taxi] ride home should the planned ride home not be available due to unforeseen circumstances.

Car park management

Some 54% of staff have parking permits and these tend to be 'front-line' staff involved in patient care, disabled staff and those required by contract to have use of a car. Claims for permits on the grounds of travel during the course of work are checked against mileage claims and evidence of need. For those with occasional need to bring a car on site (for example, when bringing in heavy equipment) there are one-day permits.

In 1997–98, a 20p a day charge was made for car parking on site on weekdays. This was increased in 1999 to 50p a day. Staff can choose to have charges taken from their salary or pay in coins on each occasion. Annual charges are made on the basis of 252 working days minus four weeks leave and two weeks for sick leave. Weekend parking has remained free. Night staff, weekend staff, disabled staff, volunteers, car sharers and tenants of the site's residential accommodation are currently permitted to park their cars on site free of charge. Parking permits are not required out of hours (though medical shifts begin or end at times when permits are required).

There is a financial incentive for staff to return their parking permits: this cash-out scheme gives staff who drive to work on at least three days a week, £250 for surrender of a permit. This incentive has been on offer since June 2000 and seven permits had been surrendered by October 2001.

The trust operates an appeals procedure in which the Director of Facilities is the final arbiter. Any changes in charges or benefits arising from the travel plan have to be approved by a Joint Staff Committee.

Other strategies

Since 1997–98 all applicants for posts at Derriford Hospital have received an applicants' pack, which contains information about the travel plan and the parking constraints, which might help with location choice for new employees who are considering moving into the area. There are moves to promote more flexible working to help reduce travel demands on the site. The Facilities Directorate has negotiated with departments in offering up to £450 to help existing car driving staff to work from home on some days, although this has not yet proved attractive to departments.

The trust does provide personalised travel planning advice to staff on request regarding public transport and car sharing.

Communications

Since 1997 the trust has communicated with staff on a continual basis about travel plan developments through a variety of media. The mechanisms for this communication have been partly through posters and newsletters.

Between 1997 and 2000 the trust produced the Derriford Newsletter to inform staff about the travel plan. Since June 2000 specific mode newsletters have been produced. There is discussion with staff consultation bodies through a Joint Staff Committee, which meets quarterly or when needed. The group is comprised of five union representatives and two or three managers.

The city council has provided some posters and the bus companies bus travel literature including bus maps. There is an annual bus road show and in 2001 a road safety show during road safety week run by the city council road safety team. This was particularly popular in covering issues such as child car seats and injuries caused to pedestrians and cyclists on the roads.

The trust works closely with the city council and acts as a lead on travel planning for other employers. The liaison officer for the trust at Plymouth City Council is the trust's previous Transport and Environment Manager, which makes collaboration and understanding much easier than it might otherwise be. There are quarterly meetings with the city council and the public transport operators at which, for part of the meeting, trust staff can ask questions about services.

The trust has made use of Geographical Information Systems to target staff living close to bus routes or where there is potential for car sharing. Letters are then sent to the specific staff members about the options available to them.

Views of those managing and implementing the plan

According to the Transport and Environment Manager, the trust suffered from a lack of available experience when it started its travel plan. There was no advice available at the time. The trust did receive original advice in 1996 from a consultant and has since maintained good links with Transport 2000, contacted similar organisations, looked at student materials and gathered what information it could. In 1996, however, no public sector organisation the trust knew of had been refused planning permission and so this was unexpected. Yet the refusal marked a turning point in that the trust was resigned to developing a travel plan.

There were some early backlashes, including a junior doctors' motion of no confidence in the hospital management, but this settled down after a while and by 1998 the local paper started to respond positively

(a) (b) (c)

Figure 3.8 The sorts of measures used at Derriford Hospital have been applied by many other employers: (a) special car sharer spaces at Boots, Beeston; (b) company-supported bus service at Orange, Bristol; (c) modern, secure cycle parking at The Open University

towards the travel plan. Nonetheless, to implement a successful strategy it is important to be able to communicate and to have charismatic managers, and to be able to manage behavioural change. This requires being 'thick skinned' and having motivation and ongoing support from within and outside the organisation. It is also important to understand that a travel plan is a living document and has to be regularly updated.

The two greatest successes have been the increase in bus use and in car sharing for which there have been high levels of support from staff. The results of these have been to reduce congestion on arterial routes into the hospital, support public transport and promote a choice of modes.

In developing the travel plan it has been important to have a good relationship with the city council, especially the person working on travel plans. Plymouth City Council has been very supportive, and has underwritten some bus routes, and put in their own funds. It has similarly been important to have good working relationships with the bus operators and to make the business case for services.

DfT, 2002, pp. 108–13

Nottingham City Hospital NHS Trust

Nottingham City Hospital NHS Trust is another example of a more established travel plan. The Trust's activities are centred on a large edge-of-town hospital site about six miles from Nottingham city centre. In 2002 it was estimated that 12 000 vehicles a day entered the site; there were 1200 parking spaces for staff and 450 for patients and visitors; the hospital employed 5200 full- and part-time staff; it had 250 000 outpatient appointments and treated 75 000 inpatient and day cases. Like Plymouth, the Trust entered a 'Section 106' planning agreement in 1997 to develop a travel plan. Ring-fenced funding from car-parking charges financed better pedestrian and cycling provision and measures to enhance public transport. Travel surveys revealed that between 1997 and 2000 driver-only ('solo') car use declined from 72% to 55%, while car sharing rose from 2% to 11%, and bus use increased from 11% to 19%.

BOX 3.5 Management of the Nottingham City Hospital travel plan

Introduction and reasons for the travel plan

In 1996 the [Nottingham City Hospital NHS Trust] provided free parking and had an unknown number of vehicles entering the site. There was little security on site and between 70–80 vehicles per month were subjected to car crime. No public transport entered the site, there was little understanding of pedestrian requirements, and there was one dilapidated cycle shed. Unrestrained car use had resulted in gridlock on site at peak times, parking chaos, and little faith in security. Additionally, the trust was entering into a Section 106 planning agreement for the construction of new buildings on the site and needed to have a co-ordinated approach to travel planning.

It is seen as essential for the trust to have a coherent travel plan in order that support facilities such as car parking are adequate to enhance the 'patient experience'. Therefore the trust has the following objectives:

- to develop a strategy for the future (2001–2006)
- a menu based approach, whereby it allows the trust board to tailor the proposals to best meet service requirements
- to provide sustainable alternatives of transportation to and from the hospital
- to ensure patients and visitors receive a quality service.

Coordination and management of the travel plan

The trust produced a first travel plan in 1997. This involved negotiations with staff and their representatives and feedback was that any travel plan funds generated from parking revenue had to be ring-fenced for transport improvements.

There is management support for the travel plan and from autumn 1996 it was incorporated into the corporate strategy for the trust. There are examples of management leading by example through the returning of managers' parking permits on account of the high frequency and low cost of bus services to the city centre. The Chief Executive has also given his personal approval to the travel plan and there is endorsement by the trust board.

Funding

Funding for the travel plan comes from ring-fencing of the parking charges. The annual cost of funding the travel plan is approximately £144 000. This is comprised of £100 000 capital to spend from car parking revenues once payments have been made for park and ride, parking management (contracted to outside service) including parking wardens, and CCTV cameras. A sum of £15 000 is spent on cycling each year and approximately £29 000 is staff costs.

Travel plan measures

These are comprised of:

- Improved cycle facilities
- A car sharing scheme
- Improved public transport provision and information
- Car parking charges
- Improved security on site
- Park and ride.

Main emphasis: car parking charges and buses on site.

Travel plan effectiveness

A [comparison of 1997 and 2000 staff travel surveys indicated that] solo car driving had reduced significantly and that bus use had increased by 73%.

Staff: Main mode of travel to/from work

Mode	November 1977%	November 2000%
Pedal cycle	5	4
Car (drive alone)	72	55
Car sharer	2	11
Bus	11	19
Train	0	1
Walk	8	9
Other	2	1
	100	100

...

Costs and benefits

As noted above, the trust spends about £144 000 a year on the travel plan. In 2001 this is comprised of:

- Car sharing £2 000
- Bus measures £8 000
- Publicity and promotion £1 500
- Cycling measures £15 000
- Staff time in managing the plan £29 000
- Walking measures £60 000
- Signage and maps £28 500

There have been some initial costs including £112 000 for the installation of CCTV cameras which was capitalised over the length of the contract. Each year the contracted parking service costs £180 000.

The annual running cost per full time equivalent employee is £41/employee (figure excludes revenue from parking).

The main benefits of the travel plan have been that it has given staff, patients and visitors a range of sustainable transport alternatives to solo car driving, together with informed choice about these options. It has been critical to put in place the 'carrots' rather than to start with 'sticks' so getting in infrastructure has been important. The most successful aspects of the travel plan have been increases in bus use and maintenance of cycle use. In particular, the introduction of buses on the site was critical in bringing about increases in bus use.

Support for bus and rail use

There is no local railway station. The main Nottingham station is located six miles away near the city centre although buses from the railway station enter the hospital site every 30 minutes.

Prior to the development of the travel plan in 1997 no public buses entered the site as services only stopped at the periphery (which is more than 400 m from the building entrances). By 2001, there were services entering the site every 15 minutes during weekdays between 7 am and 6 pm, (including between 8 am and 9 am). These services are operated by Nottingham City Transport who have funded bus shelters, a new fleet of low floor buses, and a travel map of their routes serving the hospital site. ... There are also more services that pass the hospital periphery.

There are currently no specific discounts on bus service fares generally available to employees, but this is being pursued. The cost, however, of a single ticket to the city centre at 70p makes the bus journey attractive to staff, especially as Nottingham City Council (highway authority since 1998) has introduced more bus lanes around the district. Nottingham City Transport provides a 28 day bus pass for £28.00 (£3 for initial provision of identity card) which provides unlimited travel. There is information about bus services on the hospital web site and also on the intranet for staff which have hyperlinks to Nottingham City Transport and Trent Barton Buses, the main bus service providers in Nottingham.

The trust operates a park and ride service within the site, running every 15 minutes using three minibuses, funded out of car park revenue. Two of the vehicles have been donated by the Women's Royal Voluntary Service and the hospital's League of Friends.

Support for cycling

Access to the site is reasonably good by bicycle. There have been both off and on-site improvements for cyclists since 1997. Off-site improvements include routing part of the Nottingham cycle network past the front of the hospital site. On the site, there were existing showers and changing rooms which could be used by cyclists and these are to be upgraded in 2002. In autumn 2001 there were 450 cycle stands on site. ... There are also three secure cycle compounds that can hold 90 cycles. These are remotely patrolled through CCTV cameras installed in 1998.

A Bicycle Users Group was established in 1997 although this has evolved into the alternative transport group within the hospital which focuses on all alternative modes to solo car use. The trust takes part in a range of cycling promotion events, including the annual Bike Week in June. It has a fleet of 12 bicycles for staff use and the trust pays 11p a mile for travel during the course of work. The bicycles are maintained by Raleigh (bicycle manufacturer located in Nottingham) and staff have access to lights, locks, baby seats, helmets and car racks. This is a popular service. Staff can take advantage of a 20% reduction on cycle equipment from Raleigh and a 12% reduction on the cost of a new bicycle. There is an interest free loan available for bicycle purchase.

...

Support for walking

Accessibility of the site by foot is described as 'medium'. The trust has employed consultants to advise on improvements for pedestrians in recognition that several hundred of its staff walk to work each day. A 15 mph speed limit has been introduced on the site with some cycle-friendly traffic calming measures, dropped kerbs, and new pedestrian zebra crossing installed at a cost of over £100 000. Street lighting has been upgraded and some new paths constructed.

There is also a programme of pedestrian signing being introduced or upgraded. This has arisen from an audit of the site by consultants and the development of a Pedestrian Signing Strategy in 2000.

Support for car sharing

The trust has operated a computerised matching service since June 2001. Staff can access this via the intranet and self-match. There are plans to exempt car sharers from parking charges in the revised travel plan for 2001–2006 and priority parking spaces nearer to buildings.

Car park management

There is an annual car parking charge for staff of £55.00. Each staff member can apply for a permit allowing access to the site. Staff car parking charges were introduced in 1997 at £50 a year and raised in 1999. There is a window sticker permit. As of September 2001 all students are banned from bringing cars on to the site. Currently some staff with peripatetic work patterns, such as some surgeons and community nurses who work off-site are guaranteed a parking space. In 1997 this was 3.8% of staff and the percentage has reduced to 3.2% in 2001 against a background of stable staff numbers.

There are currently 1200 car parking spaces dedicated for staff use, with 4000 'live' car parking permits issued. This results in as many as 200 staff cars parked on an unofficial basis each day.

Because of substantive improvements in car park security arising from travel plan measures the trust has received ten car parking awards since 1997. There is a trust security working group which in 2001 has been evaluating how other large organisations manage their security issues in order that an integrated system can be developed at the hospital site.

...

Communications

Since 1997 the trust has communicated with staff on a continual basis about travel plan developments. The mechanisms for this communication have been through the hospital newsletter City Post, road shows about the travel plan, articles for the hospital notice board, information included in pay packets, and emails. The trust alternative transport group has been exploring the idea of a logo to give the travel plan a unique identity.

Views of those managing and implementing the plan

According to the Environmental Services Manager, it is important to expect some bad publicity and to have a 'thick skin'. It is however, important to get communications right and so to keep the media well informed, including the internal public relations staff and the local media who will always be quick to publicise perceived opposition. At the trust, the staff and their representative wanted to see clear evidence that money raised through parking was being reinvested in transport security measures. Continual liaison with staff groups and use of internal communications media was, therefore, important. In addition, there must be support from the highest levels of management for the travel plan.

The trust has had some very good publicity from its work on the travel plan, nationally and internationally. It has also developed a very good working relationship with Nottingham City Council since 1997, with whom it had little contact prior to this time. Similarly it has developed good working relationships with the local bus companies.

In terms of plans for the future, a key objective is the implementation of a new travel plan with restrictions on parking according to distance from home to hospital. Subject to further negotiations with staff, those new staff living less than 800 metres [away] will be barred from driving to work unless they have special justifiable reasons for the use of a car. The cost of permits is to be structured so that those living closest to the hospital will pay most for a car parking permit. Staff working shifts or on rotas will be given higher priority for permits as well as those who car share.

The trust also wishes to increase bus services further and to develop a transport hub within the site. This would provide:

- a focal point for public transport
- travel information
- toilet/baby changing facilities
- facilities to meet special needs
- snacks and beverages
- travel ticket issue.

These facilities would enhance in essence the government's initiatives (The NHS Plan) of providing patients with focal points for information on a personal level.

The new travel plan will have a range of targets to be achieved. These are set out below:

Criteria	From:	To:	Date
Monitor air quality and vehicle count	Ongoing	Ongoing	May 2001
Increase disabled car parking spaces	90	150	April 2003
Increase bus use	19%	21%	April 2004
Increase cycling use	4%	5%	April 2003
Reduce single car occupancy	55%	50%	April 2005
Establish car share database	May 2001	May 2004	May 2004
Develop car parking facilities	December 2001	December 2002	December 2002
Develop travel hub/Light Rapid Transit	May 2001	May 2002	May 2002
Reduce daytime deliveries by fuelled vehicles	May 2001	May 2002	May 2002
Increase patient parking facilities	480	600	May 2003
Undertake travel survey	Bi-annually		May 2002 2004 & 2006

DfT, 2002, pp. 78–82

Overall, the performance of travel plans in hospitals has shown that, once transport management is accepted as a legitimate function of an institution, it can be applied effectively and efficiently. As noted above, at Nottingham City Hospital driver-only car use dropped from 72% to 55%, with rises in bus use and car sharing particularly noticeable. At Plymouth, over a comparable period of time, the drop in car use was from 78% to 54% – an achievement remarkably close to Nottingham's, using similar measures. Other hospitals have also achieved comparable results from their travel plans. At Addenbrooke's NHS Trust in Cambridge the drop in car use between 1993 and 1999 was from 74% to 60%. Here, as well as bus use rising from 4% to 12%, cycle use rose from an already healthy 17% to 21% (DfT, 2002).

Figure 3.9 Covered cycle parking at Addenbrooke's Hospital, Cambridge, where a travel plan has been introduced

3.6 Rising from the 'bed of nails'

The case studies above show that the use of mobility management measures allows institutions to achieve quick results in cutting energy use and emissions. But these measures do need to be tailored to the institution's needs and often require a regulatory kick-start (in these cases a planning requirement). By its very nature, transport policy is frequently subject to disagreement and controversy. The active involvement of institutions such as hospitals in developing mobility management strategies that meet their own needs could mark an important step towards a more consensual partnership approach. This may be one way to rise from the 'bed of nails'.

The next chapter examines travel plans more generally and how they are starting to be used by a variety of private and public sector organisations.

References

Cairns, S., Sloman, L., Newson, C., Anable, J., Kirkbride, A. and Goodwin, P. (2004) *Smarter Choices – Changing the Way We Travel*, UCL, Robert Gordon University and Eco-Logica. Final report to Department for Transport, London.

Coulliard, L. (2002) 'Mobility management in the Montréal region: partnership strategies and transportation management areas', Economic Community Workshop, *Proceedings for the European Conference on Mobility Management*, 15–17 May, Ghent.

Department for Transport (DfT) (2002) *Making Travel Plans Work: Case Study Summaries*, London, The Stationery Office; also available online at: http://www.dft.gov.uk/stellent/groups/dft_control/documents/contentservertemplate/dft_index.hcst?n=13896& l=4 [Accessed 29 August 2006].

Department for Transport (DfT) (2004) *White Paper: The Future of Transport: a network for 2030*, London, Department for Transport, Cmd 6234.

Department for Transport (DfT) (2005) *Transport Statistics Bulletin – National Travel Survey 2004*, London, Department for Transport.

Department of the Environment, Transport and the Regions (DETR) (1998) *New Deal for Transport: Better for Everyone*, (White Paper), London, The Stationery Office.

Department of Health (DoH), Health Inequalities Unit (2004) *Accessibility Planning: An Introduction for the NHS*, London, DoH, September.

Department of Transport (DoT) (1996) 'Transport Secretary Urges Hospitals to Reduce Reliance on the Car', press release No. 291/96, 17 September, DoT.

Dublin Transportation Office, Kirklees Metropolitan Council and the Irish Energy Centre (2001) *Impacts Calculator, The Route to Sustainable Commuting: An Employers' Guide to Mobility Management Plans*, Way to Go Research Project, European Commission SAVE II Programme, Brussels, European Commission.

Energy Efficiency Best Practice Programme (2001) *A Travel Plan Resource Pack for Employers*, Energy Efficiency Best Practice Programme, London, The Stationery Office; also available online at: http://www.dft.gov.uk/stellent/groups/dft_susttravel/documents/page/dft_susttravel_504132.pdf [Accessed 29 August 2006].

Flowerdew, A. D. J. (1993) *Urban Traffic Congestion in Europe: Road Pricing and Public Transport Finance*, London, Economist Intelligence Unit.

Freund, P. and Martin, G. (1993) *The Ecology of the Automobile*, Montreal, Black Rose Books.

IBI Group (1999) *Tax Exempt Status for Employer Provided Transit Benefits*, Final Report to the Canadian National Climate Change Process, Transportation Issue Table, chaired by Transport Canada, Ottawa, IBI Group.

Litman, T. (2001) 'Commute trip reduction (CTR): programs that encourage employees to use efficient commute options', *TDM Encyclopedia*, Victoria, Canada, Victoria Transport Policy Institute, http://www.vtpi.org [Accessed 29 August 2006].

MOST (2001) 'A ride around Rome: implementation of the Italian MM decree, mobility centres and mobility consulting', *MOST News*, no. 3, December, p. 3. Also available at http://mo.st/public/reports/most_news3.zip [Accessed 29 August 2006].

NHS Estates (2001) *Sustainable Development in the NHS*, London, The Stationery Office.

NHS Estates (2002) *New Environmental Strategy for the National Health Service*, London, The Stationery Office.

NHS Estates (2006) *Sustainable Development: Transport*, http://www.dh.gov.uk/PolicyAndGuidance/OrganisationPolicy/EstatesAndFacilitiesManagement/SustainableDevelopment/SustainableDevelopmentArticle/fs/en?CONTENT_ID=4119604& chk=m9rINj [Accessed 29 August 2006].

Potter, S., Enoch, M. and Rye, T. (2003) 'Economic instruments and traffic restraint' in Hine, J. and Preston, J. (eds) *Integrated Futures and Transport Choices: UK Transport Policy beyond the 1998 White Paper and Transport Acts*, Aldershot, Ashgate, Chapter 16, pp. 287–304.

Potter, S., Enoch, M., Rye, T., Black, C. and Ubbels, B. (2006) 'Tax treatment of employer commuting support: an international review' *Transport Reviews*, vol. 26, no. 2, pp. 221–37.

Rye, T. (2002) 'Travel plans: do they work?', *Transport Policy*, vol. 9, no. 4, October, pp. 287–98.

Schippani, R. (2002) 'Mobility consulting for companies through the City of Linz', Economic Community Workshop, *Proceedings for the European Conference on Mobility Management*, 15–17 May, Ghent, ECOMM (on CD-ROM).

Steer Davies Gleave (n.d.) *Travel Blending*, London, Steer Davies Gleave.

Chapter 4

Travel planning

by Marcus Enoch and Stephen Potter

4.1 Introduction

Chapter 3 introduced the general concept of travel plans and looked at some examples of their use by hospitals. In this chapter, the evolution of travel plans thus far in the UK is examined, together with the motivation and benefits behind their adoption by various organisations. Next, the process of developing a travel plan is explored in more detail and some of the available instruments are set out. Finally, the implications for future policy are explored.

4.2 Institutions and mobility management

Travel plans were first introduced in the mid-1990s by UK employers, and have often originated not through strategic or corporate planning, but from ad hoc initiatives in response to a particular need. For example, many initiatives that may now be labelled as a travel plan were a response to a planning requirement or a parking problem. However, 'best practice' organisations – those against which others benchmark – have adopted a more comprehensive approach and discovered that travel plans can be justified in terms of improved efficiency and can yield cost saving benefits across the organisation. The best travel plans have been integrated with actions to clean vehicle emissions and cut energy use, such as introducing cleaner-fuelled vehicles into company car or delivery fleets. Thus some public and private sector employers have repositioned their travel plans at the strategic level. This strategic approach has led to partnerships with local authorities, transport operators, other organisations and national government that have helped tackle bigger transport issues beyond the control of an individual employer. However, despite a decade in travel plan development, such best practice examples are still rare. In general, travel plans are, at most, only at the fringes of an organisation's agenda. Institutions are at the early stages of accepting that management of staff, customer and visitor travel is their responsibility. There are thus major cultural and institutional barriers to travel plans, which will take some time to overcome.

Because travel plans are an emerging transport policy response, and depend so much on institutions themselves accepting that they have a role to play, this chapter will look in some detail at the process of planning and implementing a travel plan.

4.3 Motivations for and benefits of travel plans

From a public policy perspective, travel plans are attractive to regional and local government because they are reasonably quick to introduce, relatively cheap and they are usually politically acceptable. From a company viewpoint there are circumstances where some pressing motivation (such as access issues, a shortage of parking, a lack of space or money, problems with neighbouring organisations, need for planning permission or need to enhance the organisation's image – perhaps for Corporate Social Responsibility and/or marketing reasons) means that there are potentially significant benefits for adopting a travel plan.

For example, in Buckinghamshire in the UK, the county council's travel plan has enabled it to give up about 100 parking spaces in the multi-storey car park adjacent to County Hall, at a saving of £75 000 a year (Cairns et al., 2004).

Table 4.1 shows the wider benefits obtained by organisations that have adopted travel plans.

Table 4.1 Wider benefits of travel plans for organisations

Benefits of a travel plan for organisations
Increases in bus use and associated ticket revenue
Increases in walking and cycling, with associated health gains
Improved social inclusion
Better conditions for employees
Improved staff recruitment and retention
Good public relations (PR) for businesses
May assist in meeting environmental management standards such as ISO 14001
Financial savings
Better estate management
Less noise, congestion and pollution, and better conditions for freight distribution, associated with reductions in car use
Better security and less fear of crime from better car parking management.

Source: adapted from Cairns et al., 2004.

Research into the costs and benefits of good travel plans (DfT, 2002a) showed an average cost for a travel plan as £47 per annum per employee, compared with a cost for providing car parking of at least £500 a year. Several companies, after developing travel plans, have recognised that it is something they should have done before simply because it made good business sense.

However, perhaps the key problem with tackling the transport impacts of an organisation is that any policy initiative tends to be viewed as an externally imposed regulation. In such a case, the usual response is to find the cheapest way to achieve compliance and leave it at that. It is thus not surprising that when travel plans are introduced as a result of a planning condition, or a strategic NHS requirement upon existing hospitals, they come to be viewed in this way.

This in itself is a serious problem, since the idea that there can be *benefits* in managing the travel of staff, customers and visitors is rarely acknowledged. This is probably the most fundamental barrier facing the development of travel plans in the UK, and indeed in other countries where similar programmes have been introduced.

Figure 4.1 The staff commuter centre at Park Royal business park in London. Here, the travel plan is marketed as a service to employees on the site

However, these gains to an organisation tend to be spread across several departments and are not immediately evident or specifically identifiable, while there are also institutional and cultural issues that make it acceptable for organisations to provide car parking but not to fund public transport tickets or cycle training schemes. Thus, in the absence of the type of motivation described earlier, most organisations have simply not participated in helping to solve something that is not legally or institutionally 'their problem'.

A number of studies (Bradshaw, 1997; Coleman, 2000; Rye, 2002) report that less than 10% of large private businesses (of over 100 employees) have adopted travel plans, while small- and medium-sized enterprises (SMEs) have taken even less of an interest. This lack of interest is for a number of reasons. In particular, Rye (2002) identifies a number of key barriers to wider travel plan implementation, namely:

- companies' self-interest and internal organisational barriers
- lack of regulatory requirements for travel plans
- personal taxation and commuting issues
- the poor quality of alternatives (particularly public transport)
- lack of examples due to novelty of the concept.

In addition, while the UK Government has formally recognised the travel plan since its inclusion in the 1998 White Paper *A New Deal for Transport: Better for Everyone* (DETR, 1998), and has provided a whole series of support measures, these have tended to be rather small scale, incremental and randomly applied. Travel plan policy, meanwhile, has largely been reactive and somewhat lacking in an overall strategic direction.

Despite these barriers, though, travel plans have somehow survived and over the last few years have begun to make an increasing impression on the

Residential travel plans shift the emphasis to the journey origin and instead aim to provide transport options to a range of possible destinations (e.g. work, education, shopping and leisure). In the UK, while such plans have only really taken off with the publication of guidance from the Department for Transport in late 2005 (DfT, 2005), the BedZED scheme is one much earlier example of how such a travel plan can be applied (Morris, 2005) (see Box 4.4).

BOX 4.4 BedZED: An early residential travel plan

The Beddington Zero-Emission Development (BedZED) is located in the London Borough of Sutton. It has 82 homes, 16 business units, childcare and community sports facilities. The site is on a bus route. It is 5 minutes walk from a rail station and 10 minutes walk from a tram stop. The project incorporates energy-efficient design, recycled materials and a combined heat and power plant. The area is not within a controlled parking zone, so no mechanism exists for enforcing off-site parking. In order to address concerns of overspill parking, a Green Lifestyles Officer was employed to establish a car club and Green Transport Plan. As a result, the parking standard was reduced by 50%. The site is defined as low-car. Nevertheless, it has a parking standard of 0.85 (84 spaces), which represents a considerable proportion of the available area.

Morris, 2005

Travel plans in the UK tend to be associated with measures to cut the transport impacts of individuals journeying to and from a site. This could be in relation to employees, schoolchildren, customers, students or football fans heading to a stadium (yes, several UK football clubs have 'fan travel plans'). But the transport of goods and deliveries in general can also account for a substantial part of the transport impacts of an institution. Mobility management can be applied to freight as well as to people. The idea of Quality Freight Partnerships – focusing on goods delivery and distribution issues rather than on people only – are now gaining acceptance (see Box 4.5).

BOX 4.5 Freight travel plan: Denholm Industrial Services

Road freight haulage company Denholm Industrial Services is a good example of how a freight travel plan can operate successfully (DfT, 2003). This was adopted in 2001 and drew on expert advice which suggested improving monitoring and targeting vehicle miles per gallon, mileage and fuel costs, as well as reviewing vehicle utilisation and vehicle specification and payload performance. This analysis highlighted areas of inefficiency and resulted in annual fuel economy savings of £36 000 (on a fuel budget of around £200 000); 150 000 fewer miles travelled; and improved productivity.

Other freight travel plans are also now being adopted, often by local authorities. Thus, Bristol City Council has established a Freight Quality Partnership and is investigating the potential for freight transhipment centres, whereby freight from large vehicles is decanted into smaller vehicles on the outskirts of the city for delivery in specific neighbourhoods.

Scope

A second major trend that has occurred has been in the scope of travel plans. In particular, while the first plans were applied by the organisation themselves to mitigate existing problems, by the late 1990s a number of local planning authorities were beginning to make the link between travel plans and planning consent. Therefore, by 2001 a survey for the Department of Transport, Local Government and the Regions found that 156 local authorities out of 388 surveyed required the developers of some proposed developments to set up a travel plan as a condition for being awarded planning permission (DTLR, 2001b). However, until the changing of planning guidance in 2005 with the issuing of Planning Circular 5/05 (ODPM, 2005), such rules and regulations tended to be made on a case-by-case basis with no guarantee that an effective plan would be in place following the results of the negotiation phase. With the new guidance, though, local authorities are now encouraged to develop standardised, transparent, and area-based approaches to planning decisions, and in London this has had significant ramifications. Here, Transport for London (TfL), the capital's transport authority, is currently in the process of drawing up guidance for London Boroughs that aims to ensure that some form of travel plan will need to be provided for every planning application submitted in the capital.

There is also evidence that the scope of travel plans has been extended to more existing organisations. For example, all NHS facilities and all government department offices have been required to adopt a travel plan for a number of years, while some commercial organisations are applying similar regulations based on internal drivers (typically driven by cost saving and/or by corporate responsibility agendas).

Scale

Meanwhile a third trend that has started to emerge since the beginning of 2005 is the development of so-called local travel plan groups or networks. These have come about for a number of reasons, but fundamentally these are that:

1. Groups are collectively able to achieve more than single agencies or employers when dealing with common concerns (thanks to pooled resources delivering higher investment, dedicated staff, and greater political influence) and yet allow the member companies/organisations to focus more on their core competencies.
2. Groups have the ability to move Transport Demand Measures (TDM) from a site-specific application to more flexible and effective area-wide application.
3. Groups can improve the level of communication between the sectors and allow the level of flexibility necessary to ensure that transport objectives are met in ways that maximise the benefits for businesses, residents and commuters.

Enoch, Zhang and Morris (2005) provide an overview of the various types of group in place as of mid-2005 and develop a basic framework to classify their structures and functions. What is particularly interesting is that some of these groups began as quite informal networks (see Box 4.6); but are now following the trend seen in North America and becoming increasingly formal. Some also include not only business organisations, but residential areas and shopping facilities too – for example, the Dyce Transportation Management Organisation in Aberdeen (see Box 4.7).

BOX 4.6 **Temple Quay Employer Group, Bristol**

Bristol City Council set up a Green Commuter Club in 1999, following a conference designed to promote travel plans among companies in the city. This now has more than 85 members and meets on a quarterly basis.

In 2001, a group of the members were about to move into a new development area known as Temple Quay and so decided, together with the City Council, to set up their own subgroup. The Temple Quay Employer Group now has 15 members both in and next to the newly developed area. These include Orange, Norwich Union, Highways Agency, DEFRA and Bristol and West.

Members of the subgroup are required to sign up to a statement of intent which commits the company to addressing common issues. Projects – e.g. a car sharing database – are financed by contributions from the Council and member companies on a project by project basis. Initially, the TQEG was run by the council, but recently some of the organisational effort has been taken on by Norwich Union.

Bristol City Council has since tried to establish a second sub group in the Avonmouth area of the city, but this has struggled to attract much interest probably due to the area's relative inaccessibility by alternative modes to the car. Most recently, the council is examining the feasibility of establishing a third group in the South Bristol area.

Sources: Ginger, 2005; Cairns et al., 2004

BOX 4.7 **Dyce Transportation Management Organisation, Aberdeen, UK**

The Dyce Transportation Management Organisation in Aberdeen, Scotland is thought to be the first such organisation in the UK, whereby a diverse group of businesses come together specifically to address travel issues within a more formalised relationship (although of course there have been several business parks that have established travel plans). Dyce is an area of 20 000–30 000 commuters with mix of employer types between the docks and the airport. The public transport access is poor. With support from Aberdeen and Grampian Chamber of Commerce, and consultant Vipre, a not-for-profit organisation was established, called 'Dyce Transportation Management Organisation'.

All businesses in the Dyce/Kirkhill/Stoneywood area were invited to join, irrespective of their number of employees. The initial TMO group includes several companies involved in oil exploration, such as BP and Halliburton. Between them, these members employ over 3000 onshore staff.

The idea of setting up a TMO came up shortly after mobility management consultant Vipre approached BP (and later the council) in March 2004. In July 2004, the regional body NESTRANS (the North East Scotland Regional Transport Partnership) appointed a Travel Awareness Project Leader, whose role was to promote travel awareness, encourage more businesses and other organisations in the region to adopt company travel plans and to reduce dependency on private cars. Public money (£70 000 from NESTRANS and the Council) was used as seed funding. Of this, a one off total of £20 000 was invested in producing relevant materials, conducting surveys etc, while the remainder was to be spent on the TMO management at around £4000 a month. This was used for the background research, set up costs and the cost of employing a project manager for the first six months. Organisation members will also make a contribution to the management cost through an agreed funding formula (£10 per employee per year). This is because payment of a membership fee means that the organisations are more likely to take the plan seriously and expect results. From financial year 2005–6, 50 per cent of the operational costs were to come from TMO members.

Research began in August 2004 when Vipre conducted an online survey within the area and successfully got back 2500 responses from local businesses in the following month. Aberdeen City Council then identified a set of travel plan measures including car sharing, van pooling, public transport operator network changes and so forth. A travel plan was finally set up around late-November 2004.

As for the performance of the TMO, Aberdeen City Council has said it will be measuring a number of indicators including number of people car sharing and the number of people driving on an annual basis.

Sources: Aberdeen City Council, 2004; Murphy, 2005; NESTRANS, 2004; Caswell, 2005

In addition to the trends directly affecting travel plans and the transport agenda, it is also clear that such a shift towards this neighbourhood-based model of service delivery is not just confined to the transport sector – for instance, policing and healthcare have been moving to such a devolved model for a number of years. Until now though, transport has usually been an absent voice even in such policies as the Sustainable Communities programme, run by the Department for Communities and Local Government.

Travel plans are gradually becoming a more embedded part of transport practice and policy and so increasingly recognised throughout society. The next section will present how travel plans are developed and what measures and instruments are available to travel planners.

4.4 Developing a travel plan

When an organisation is developing a travel plan, it is important to take into account the specific needs and circumstances of each site. In particular, travel plan development depends on such factors as organisation size, location, the nature of the business (which influences the amount of business travel, number of visitors, number of deliveries, etc.), the reasons why a travel plan is being developed, staff attitudes towards different measures, and the resources available. For example, a travel plan for a large retailer's depot, which has a considerable amount of heavy goods vehicle (HGV) traffic and where the workforce all live locally, will be different from a travel plan for the national headquarters of a high-tech electronics company, where commuting is the main transport impact and most employees live further away. Finally, a travel plan is a continuous programme, and must be resourced, maintained and reconsidered periodically as the requirements of an organisation change.

There are a number of stages that organisations need to go through when introducing a travel plan. The following breakdown of this process into seven steps is taken from the Cheshire County Council's *Commuter Plans in Cheshire: Steps to Success* (Cheshire County Council Travelwise, 2002), with 'commuter plan' replaced by 'travel plan'. The latest version of this report was issued in 2006 (Cheshire County Council Travelwise, 2006) and offers similar advice on best practice for producing effective travel plans.

Steps to success

Depending on a company or site's particular needs and objectives, the travel plan may form part of a wider company transport plan. This can cover a wide range of issues including commuting, business travel, [visitor and customer travel], fleet management [including alternative fuels], deliveries and other commercial activity.

Step 1: Identify the problems and make the case for action

As already highlighted, there are many factors which can influence the decision to adopt such a strategy. [Some of these may] include:

- Concern about the impact of traffic congestion ...
- Pressures of on-site car parking demand ...
- Expansion plans leading to significant on-site development ...
- Environmental considerations ...
- Leading by example ...
- Conditions of planning consent ...

...

Step 2: Secure commitment and allocate resources

It is essential that all Directors and Senior Managers understand and support the aims and objectives of the travel plan and are prepared to lead by example. This is vital to win over staff and Trade Union support and co-operation. [It is notable that this was one of the factors considered to be of importance in the hospital examples in the previous chapter; also see Box 4.8]

- Set a challenge to your senior managers to attempt to reach the workplace without using their cars.
- At least one senior manager should sit on, or chair, a steering group responsible for guiding the project's development.
- Appoint a dedicated travel plan/staff travel co-ordinator to lead the project's development.
- Secure funding to support the successful development of your travel plan.

Successfully introducing a travel plan can be greatly assisted by a variety of measures that can help send the right messages to all employees about the company's commitment to their particular programme of activity. Suggestions include:

- Review your existing company culture. Does this act to encourage car use? Can steps be taken to promote a more sustainable approach to travel?
- Consider your senior managers and directors giving up reserved parking spaces and pledging to use alternatives to their own cars whenever appropriate, to lead by example.
- Review your car park management and entitlement to parking permits. It may be suitable to revise allocation with priority given to work related needs.
- Ensure that all maps and guides showing your company's location for visitors and clients include details of how to reach the site by public transport and cycle.
- Ensure that your company includes staff travel information in new starter induction packs. This may include local public transport timetables or a registration form for a car sharing scheme.

Step 3: Raise awareness and build consensus with employees

The key to success is staff 'ownership' and involvement. Your employees will need to be informed regularly about what the travel plan is trying to achieve, how it is doing this and, most importantly, what benefits will be gained by individual employees as well as the company as a whole.

Boxes 4.8 and 4.9 and the intermediate text, which are edited extracts of the DfT's *Travel Plan Resource Pack for Employers* (2002), explore the softer issues in travel plans, especially in relation to Steps 2 and 3 of *Steps to Success*.

BOX 4.8 Travel plans: working with human resources staff and trade unions

Steps 2 and 3 raise the issue of the changes involved in implementing a travel plan and the fact that these may not be welcomed by some employees and users within an organisation, even if it is for the benefit of all. This crucial issue is addressed in the DfT's *Travel Plan Resource Pack for Employers* (2002), from which the edited extract in the remainder of this box is taken.

A successful travel plan will need the support and commitment of all members of your organisation. Travel plans can have an impact on conditions of service and, in some instances, staff may interpret the proposed changes as an attempt to reduce their current benefits. It is, therefore, important to involve the Human Resources (HR) department and Trade Union (TU) representatives at the earliest possible stage.

Getting Union and HR staff 'on side'

...

Developing and implementing a travel plan is a two-way process. Your organisation is trying to make more travel options available to staff but in return expects employees to take up the options, at least some of the time. Much depends on goodwill. ...

> Involving Union and HR staff early in the process will demonstrate that their input is valued. Once involved, they need to be kept informed and an open channel of communication maintained. Clarity and openness about plans will also help to maintain goodwill.
>
> Attitudes towards your plan will depend on your organisation's circumstances and ethos but also on the personal views of individual staff members.
>
> Energy Efficiency Best Practice Programme, 2002, Sections 2.10 and 2.11

The general arguments in favour of developing a travel plan for your organisation will be relevant for all, but some arguments will carry more weight than others, depending on the individual.

... It is likely that the Union will have a policy on transport issues. Find out more about the stance taken by the Union(s) in your organisation. It is possible that the local representatives may not be fully informed and if you can demonstrate that the principles behind your proposals are in line with Union thinking they may be more supportive ...

Importance of having Union and HR staff 'on side'

...

... Union and HR staff can be involved in various ways [including: data gathering, ideas development and implementation, for example]:

...

[HR staff can be actively involved in the following tasks:]

- Include a travel plan briefing for new employees in the recruitment and interview process.
- Advise on developments elsewhere in the company which could have an impact on the travel plan.

[HR and TU staff can be actively involved in the following tasks:]

- Help to disseminate and conduct the survey.
- Help to ensure the suggestions are fair and realistic.
- Be an initial sounding board for new ideas before they go out to consultation.
- Assist in organising consultation, focus groups, feedback sessions.
- Suggest incentives for take-up.
- Ensure initiatives from 'grass roots' are driven forwards (bottom up ideas are often more likely to be accepted by the workforce).
- Raise awareness of why the travel plan is being developed.
- Include HR/Union representatives on interview panels when appointing a travel plan co-ordinator.

[...]

BOX 4.9 Travel plans: working with organisational culture

In most cases, transport issues and modes of travel are not central to an organisation's concerns. The ease with which your organisation will adopt a travel plan will be influenced by the kind of culture already in place. Organisations with an open and accessible management style and with effective internal communication structures are likely to be good candidates.

Understanding the way your organisation functions internally, and in relation to those outside, will help to identify the most appropriate ways of introducing change. Likewise, understanding how employees see themselves in relation to the organisation will help determine suitable approaches. Do they feel engaged with the organisation and responsible for their travel choices?

Corporate culture tends to come from the top. It is, therefore, necessary to ensure the most senior executives support the change.

Plan the change

Change is resisted if there is no perceived need for it. A request to complete a Staff Travel Survey can be the first employees hear about impending change. This will often be too late. It is very hard to sell something to someone who does not know they might want it. So, start the debate two or three months before your survey is undertaken. This can be done very overtly with a poster campaign around the premises. This could start with some national statistics and progress to narrower, local facts. These could include statistics on time spent commuting to work ..., costs of commuting by car ... and environmental information ... Include facts specific to your organisation.

Discussion can also be initiated through line management. Use the members of your travel plan Steering/Working Groups to raise the issues in their own departments or make use of team briefings as fora for discussion. Once the ground has been prepared, rolling out the Staff Travel Surveys and site audits will ensure the issue remains in people's minds.

Bringing about change

Knowing what makes your organisation tick is crucial to identifying which measures might be appropriate to your organisation.

[...]

Answering the questions below will help you to devise an appropriate strategy.

- What is going to motivate staff?
- Are senior management supportive? ...
- Are senior management leading by example? ...
- Does your organisation already have a good internal communications network? ...

- Does your organisation have a consultative style in its decision-making processes? [Staff familiar with these kinds of processes can encourage others to contribute]
- Do you fully understand the various communications and decision-making channels in your organisation? …
- Are environmental issues already of interest to your organisation? If they are, travel issues will be readily understood to be part of that.
- Do staff see themselves as responsible for their travel mode or is it seen as 'someone else's problem'? Your travel plan will need to take attitudes and expectations into consideration.

Figure 4.2 A travel plan awareness display for staff

[…]

If you can demonstrate that your organisation is genuinely interested in providing benefits for staff along with benefits for the organisation, they are more likely to respond favourably. Success is very attractive. If staff see an initiative is successful, they will be encouraged to join in.

Energy Efficiency Best Practice Programme, 2002, Sections 2.10 and 2.11

Step 4: Gather data

Before decisions are made on what to include in your travel plan it is vital that you gather data on existing travel habits and alternatives [see Figure 4.3].

The survey will act to review 3 key issues:

- Where people live.
- How they currently travel to work.
- Their willingness to use alternative types of transport instead of their cars.

This may require quite detailed questioning to understand factors that influence existing travel patterns and the necessary measures which may encourage people to use alternative types of transport.

It is also useful to build up a company profile. This will include details of workforce size, hours worked, number of car park spaces provided and the cost of this, existing measures to encourage alternatives to the car and consideration of future expansion plans.

Figure 4.3 A commuter travel survey under way at Cambridge railway station

Step 5: Review and evaluate alternatives to the car

For many companies and their employees introducing a travel plan will mark a major break in prevailing company culture and car dependency. The preparation of the travel plan must take this into account. [This will be considered later in this section.] Your staff need to be satisfied that the proposals are not anti-car (many staff will feel that they have no real alternatives to using their car), neither should it impose the impossible or unworkable in its recommendations. Instead, it should build on information gained from the staff travel survey, particularly details about employees' willingness to switch [travel] modes and the measures required to bring this about.

...

Step 6: Agree the strategy and set targets

In setting targets, the overall aim should be to seek:

- A reduction in single occupancy vehicles accessing your site.
- To increase the use of alternative modes of transport to the car.
- To set targets. The Government's Advisory Committee on Business and the Environment recommends that companies set a 10% target to reduce the number of employees commuting to work as the single occupant of a car.
- To be realistic, success will not come overnight. However, it is not unrealistic to consider a reduction in car use of at least 10% and maybe as high as 30% over a three year period, depending on location.
- To have regular reviews will greatly assist the success of the plan.

Step 7: Make it happen and maintain momentum

Get the formal launch of your travel plan right. This will pave the way for its success ...

This launch could include a challenge to help create staff ownership of the project.

The success of the travel plan depends very much on the level of your commitment, the resources available and the perceived attractiveness of the alternatives.

Appointing a dedicated staff travel co-ordinator will prove a crucial move in bringing these activities together and helping the overall development of the travel plan.

Summary of key points

A summary of some of the key points to consider in designing and implementing a travel plan is given in Table 4.2.

Table 4.2 Summary of key points for formulating and implementing a travel plan

1	Identify key influencers in your organisation. They may not be the most senior, but they will be respected for their achievements. Enlist their enthusiasm.
2	Identify and enlist support from those that have an 'environmental conscience'.
3	Enlist the support of Union representatives if you have them.
4	Develop a marketing strategy. Use your organisation's specialist staff if you have them.
5	Develop an internal communications strategy, again enlisting specialist help if it is available.
6	Identify if there are other areas of your organisation where changes are necessary or being made. Can you work together?
7	In a large organisation, there may be one or two departments that already have a culture that will make them more amenable to change. Focusing initially on them and being able to demonstrate success there, is likely to make change easier elsewhere.
8	It is important to 'sell' the travel plan to staff [see Figure 4.2] at the recruitment stage and to get them on-board and supportive of your plan's objectives as soon as they join the organisation.
9	If you are part of a large, multi-site organisation, with national policies that affect travel and transport issues, you may need to address policy changes with Head Office personnel/management. Head Office should themselves be encouraged to develop a travel plan and lead by example.

Source: Energy Efficiency Best Practice Programme, 2002, Sections 2.10 and 2.11

4.5 Travel plan measures

The above section went through the strategic steps and issues involved in setting up and running a travel plan. In this section consideration is given to the specific measures that a travel plan can contain. As was noted at the beginning of Section 4.4, the package of travel plan measures will need to be tailored to each individual site, and the way this happens in practice has already been noted in the hospital case studies in Chapter 3. This section draws upon wider experiences of travel plans to review measures that have been used and determine how well they work in practice.

Reducing the need to travel

One obvious way of reducing parking and traffic problems is to reduce the need for making journeys in the first place. Reducing transport dependency was identified in Section 1.8 of Chapter 1 as a crucial component to cut energy and emissions from transport to a sustainable level. One travel plan measure that takes this approach is the introduction of flexible working arrangements that permit employees to travel a little earlier or later than normal to avoid the busiest time on the road, thus saving time and leading to some reductions in fuel consumption and emissions due to better driving conditions. The impact upon energy use and emissions of such practices is marginal, but other flexible working practices can have a substantial impact.

One example is 'compressed working', which may involve people working, say, four-day weeks or nine-day fortnights, but with longer days. In California, the city of Irvine introduced a compressed working week in 1991. During the first nine months, not only did this cut the amount of commuting and pressure on parking spaces, but there was also a 16% reduction in sick leave and a 17% reduction in staff overtime worked compared with the same period the previous year (DETR, 1999). This illustrates the indirect (and significant) benefits to an organisation that travel planning can achieve. Another example is BP's office at Sunbury-on-Thames, where staff are encouraged to work slightly longer days in return for a day off each fortnight. Pfizer, in Kent, also offers staff a compressed, nine-day fortnight. Pfizer also has plans to set up satellite offices in areas where staff live, overcoming potential problems of isolation for those working at home (DfT, 2002a).

Chapter 1 also looked at teleworking, whereby people work from home using communication networks. The widespread use of such flexible working practices can also permit 'hot desking', where people share workstations, rather than have one each, which may otherwise be under-utilised. Alternatively, occasional work spaces can be provided in company sub-offices. If flexible working practices reduce the amount of office space required, very substantial savings indeed can result, especially in high-cost city-centre offices.

Buckinghamshire County Council has promoted home working and 'hot desking', where possible, to reduce travel, and at the Government Office for the East Midlands in Nottingham, flexitime is encouraged and laptop computers are available for staff to use at home or on public transport. Boots is another company that has encouraged home and flexible working practices to be more widely adopted, and AstraZeneca staff can apply to have a web camera on their PC or laptop (DfT, 2002a).

Such approaches are usually popular among staff, while the costs to the employer can be minimal and can even result in large savings. However, there are implications for administration. Management is also often concerned about staff supervision, although this is viewed as a rather old-fashioned attitude. According to the *National Travel Survey 1998–99*, in Britain about one million people (or 3.7% of the workforce) usually worked from home, and a further two million used their home as a work base but also worked elsewhere (DTLR, 2001a).

One reason that people often drive to work is because they need to visit a bank or go shopping during their lunchtime, a trip that would not be possible without a car. Providing on-site services, such as a shop, chemist, newsagent or cash dispenser, can therefore help, particularly in larger, more isolated locations. Even if people still drive to work, such measures cut down on the amount of driving needed. For example, at The Open University site in Milton Keynes, a van from a local Waitrose supermarket used to deliver pre-ordered goods at the end of the working day to a car park where staff could load them straight into their cars. Some companies, particularly those in out-of-town locations, also operate free or subsidised 'works buses' for lunchtime shopping trips. A combination of these measures is implemented at the Driver and Vehicle Licensing Agency (DVLA) offices at Swansea, which have an on-site pharmacy and a dry-cleaning collection service, and also operate a lunchtime shopping shuttle bus service into the town once a week (DETR, 1999).

Reducing business travel can be integrated into a travel plan and can result in substantial cost savings to an organisation. New technologies play an important part in enabling a change in travel behaviour. Information technologies such as the Internet, teleconferencing or phone conferencing can remove the need for trips altogether and assist home workers. Several companies are beginning to expand their use of videoconferencing to cut business travel. These include Vodafone in Newbury, Egg in Derby, the Government Office for the East Midlands in Nottingham, and BP at its Sunbury-on-Thames office. AstraZeneca has set up six to eight videoconferencing studios at its site in Macclesfield (DfT, 2002a).

Encouraging travel to work by train or bus

Transfer to public transport is, as was noted in Chapter 1, widely espoused as an environmental measure. The extent to which this is viable for individuals varies greatly, as does its appropriateness for employees and visitors to specific sites. City-centre sites are likely to be better served by public transport than other areas, but the example in Chapter 3 of Derriford Hospital in Plymouth, on a suburban site, shows that it is possible to significantly improve public transport access elsewhere as well.

There are several ways in which employers can encourage their staff to get to work by public transport. For instance, they can:

- provide public transport information in the workplace
- negotiate public transport discounts from bus or rail operators to either enhance services or reduce fares
- subsidise public transport (to enhance either services or fares)

- provide 'works buses' to supplement existing public transport, or
- promote rail for business travel.

There are a number of examples of successful train and/or bus measures in travel plans (see Figure 4.4). One is Buckinghamshire County Council, which negotiated significant discounts for staff to use local public transport. As a result, staff paid half fare on Arriva buses and got a third off rail fares on services operated by Chiltern Railways. Both operators attracted enough new custom to profit from the arrangements, and public transport use among County Council employees increased from 8% to 14% (DfT, 2002a).

Stepping Hill Hospital in Stockport negotiated a discount with local bus and train operators of approximately 5% for staff displaying employee travel cards. This may be further subsidised by the hospital to give a 20–30% discount, using revenue raised from car park charges, as is done at Derriford Hospital (NHS Estates, 2001). In addition, as noted in Chapter 3 (Nottingham City Hospital), buses were diverted to the site to drop off and collect passengers.

Egg, the eBank based in Derby, introduced several measures. A public service shuttle bus, subsidised by Egg and used by 14% of staff, ran every 12 minutes between its site and Derby bus station. While the service was initially free to staff, a nominal 10p fare was introduced later. Also introduced was a free contract bus between Egg and a nearby park-and-ride site. In addition, liaison with the local authority led to the installation of two new bus stops and shelters close to site entrances (DfT, 2002a).

The mobile phone company Orange funds a fleet of six single-deck buses to operate on two routes between Aztec West and Almondsbury business parks in the north of Bristol and its new Temple Point office in the city centre.

Figure 4.4 The pharmaceutical group Pfizer operates works buses that connect its site to the local railway station (as does the mobile phone company Orange)

BAA, the operator of London's Stansted Airport, had an impact on the local buses that extended a considerable distance from its site. This was because it used its travel plan to address a problem of staff recruitment. With a shortage of staff locally, the company sought to recruit people living further away along public transport corridors, particularly the rail corridor into London. This resulted in money being spent to improve the quality of the 123 bus route that links Ilford and Wood Green in east and north London to Tottenham Hale station (the only stop on the Stansted Express service from London's Liverpool Street station). So people travelling on the 123 bus in London now have an enhanced service.

An important development is that the growth in travel plans has begun to produce initiatives from bus operators. For example, bus company First Hampshire has targeted a thousand companies in its area with details of its travel plan scheme. The *Take One to Cure Congestion* leaflet (designed to resemble a packet of aspirins) is aimed at raising awareness of the company's existing bus network, as well as explaining how it can plan and operate bespoke staff shuttle bus services for employers in the area. Similar schemes have already been established with Portsmouth City Council and with the main hospitals in both Portsmouth and Southampton (*Transit*, 2002).

Car park charges and cash-out

The provision of free or cheap car parking is, in practice, a subsidy provided by an organisation only to those who drive. Indeed, the availability of a free car parking space is one of the main influences an employer has on people's travel behaviour. Surface car park construction costs £400–£800 per space plus annual maintenance of £100–£500, while the cost of building each multi-storey or underground space is in the region of £6000 (Energy Efficiency Best Practice Programme, 2002). Consequently, introducing parking controls, restrictions and/or charges, or paying staff to give up their parking space, can be very cost effective. But the very effectiveness of charging staff for parking also often makes such actions extremely unpopular and difficult to introduce.

An alternative, which is obviously more acceptable, is to pay drivers not to use their cars for certain trips – effectively bribing motorists to use an alternative mode. One application of this idea, the 'parking cash-out', is in use in the UK.

As noted in Chapter 3, Derriford General Hospital in Plymouth has a parking cash-out scheme that applies to staff who regularly commute by car three or four days a week. Applicants are monitored over a four-week period to see if they qualify. If they do, they are then given a one-off payment of £250, plus an extra amount to cover VAT. In return, staff members forgo their right to park by handing over their parking permits and having their ID codes erased from the parking monitoring system. The scheme was introduced in mid-2000, but by 2003 only seven people of the 3500 (0.002%) who qualified for a parking permit had taken up the benefit (although 25–30 people had applied). In 1997 airport operator BAA offered employees £200 each to forgo their parking spaces at Heathrow. This was a little more successful than at Derriford Hospital, with 33 (around 1%) taking up the one-time offer.

Rather than giving just a one-off cash-out, a scheme started in 1995 at Southampton General Hospital gives car park permit holders an initial payment of £150 and subsequent annual payments of £96. Take-up is larger than at BAA or Derriford. As of autumn 2001, 551 out of 5911 permit holders (9%) had taken up the scheme. A monthly, rather than annual, system is in operation at the Vodafone offices in Newbury, Berkshire. Introduced in 2000, the scheme allows any employee to opt out of having a parking space and receive an extra £85 in their monthly pay packet. This substantial incentive has resulted in 1500 (a third) of the 4500 staff based in the town taking up the scheme.

From the above examples of parking cash-out, it appears that there is a pattern of take-up related not only to the amount of money offered, but also to the degree of flexibility involved. It is one thing to say that you will not be able to drive to work for a month and then review the situation, but quite another to say you will never drive to work again! Furthermore, if a scheme is inflexible employees who might feel happy not to drive on one, two or three days of the week cannot benefit because they need to use the car on the other days.

To address this problem, in 2001 the pharmaceutical giant Pfizer introduced a flexible parking cash-out scheme that rewards non-car commuters on a daily basis at its research and production facilities at Sandwich in Kent and at Walton Oaks near Reigate in Surrey. This works by using a staff-personalised security pass involving 'proximity card' technology (see Figure 4.5). An employee's card is credited with enough points to 'pay' for one month's parking. The card opens the parking barriers and records how many points are used. At the end of each month staff cash in any points they have not used for parking, payments being made through the payroll. Staff at the Sandwich site receive £2 per day for leaving their car at home, while at Walton Oaks the incentive is £5 per day – a reflection of the far tighter parking standards set by the local planning authority. Overall, it is estimated that the value of cash-outs given to staff will cost Pfizer around £0.5 million per year. The impact upon travel choice is, however, substantial. In 2003 around a third of staff travelled to work other than by car to locations that would normally be very car-dependent.

Figure 4.5 Using the Pfizer 'proximity card' pass to pay for parking at the company's Sandwich site. Staff can collect £2 per day if they leave their cars at home

The Pfizer example illustrates how much easier and more effective it is to persuade people to switch from using the car for one or two days a week than for four or five days a week (or forever). The travel plan measures that allow people to regain their right to park if they have a baby or move house, or ideally to park when the weather is bad and cycle when it is dry and sunny, are likely to be more appealing. This shows the care that is needed in developing a travel plan measure. Being stingy will yield little benefit for the costs incurred. Parking cash-out can be an expensive travel plan measure, but may not be as expensive as providing parking. At an annual cost for surface spaces of at least £500 and for multi-storey spaces of £6000, an annual cash-out payment of £500 can be a cost-effective alternative to car park construction. But it will also require close cooperation from a local council to control parking in streets near a site, to prevent staff from abusing the system (by taking the cash reward and then parking nearby). Nevertheless, providing financial incentives to drivers, and existing commuters who already do not drive to work, certainly generates far less staff opposition than introducing charges or just restricting parking spaces.

Walking and cycling to work

As nearly 60% of all car journeys to and from work are less than five miles long and a quarter are less than two miles, promoting walking and cycling can play a significant role in reducing car trips. Travel plan measures that can help encourage walking and cycling include:

- promoting, via staff newsletters, the health benefits of walking and cycling to work
- identifying safe routes to the workplace and publicising them in maps or guides
- providing lockers, showers, changing rooms and secure cycle parking
- negotiating with the local authority to build safer off-site walking and cycling routes
- granting interest-free loans and/or discounts for the purchase of bicycles.

The sustainable transport plan prepared by the Stepping Hill NHS Trust in Stockport included a number of measures to promote walking and cycling. Specifically, it introduced 'green' route maps and newsletters, built showers and changing facilities, and improved cycle parking security. It also negotiated discounts with local bicycle shops, offered a 'bike doctor service' on site for bicycles that needed repairing or maintaining, and mounted awareness-raising events. Finally, from the summer of 1996, the Trust purchased 85 bicycles for staff to lease, with a commitment to buy a further 25 each year. The scheme was funded from car parking revenue (Collins, undated).

Car sharing/lift sharing

In the context of travel plans, the term 'car sharing' usually refers to offering lifts to work, school or college. How this is done ranges from informal lift-sharing arrangements among friends within one business or street, to formal arrangements using computer databases. In general, employers can help staff by establishing a car-sharing database, giving parking priority to car pool vehicles (see Figure 4.6), charging car pool vehicles less to park, and setting up a guaranteed ride home scheme to cover emergency events for lift sharers. Sharing a car where people are attending the same meeting can also be promoted as a way of reducing single-occupancy car trips for business journeys; an example is the provision of a car passenger mileage rate for business trips to reward motorists who transport colleagues to business meetings.

In Milton Keynes, UK, the car sharing scheme CARSHAREMK attracted over 1000 members in its first nine months through incentives such as free parking for sharers, dedicated parking bays in prime locations in the town centre and discounts on the local buses. The scheme was launched at the same time as further charges for parking in the town centre were introduced in October 2002 (Cairns et al., 2004). It is used primarily for commuting into Central Milton Keynes (CMK). As of summer 2003 over 90% of the members routinely used the scheme, and shared cars made up nearly 8% of the total town-centre parking at peak times during the week. Recruitment was steady at about 100 new members per month, even though there had not yet been any concerted campaign to target the large employers in the town.

Over 30% of employees working at Marks & Spencer Financial Services in Chester now car share on one or more days a week. Sharers are matched using a computer database, are offered the most convenient parking spaces at the front of the building, and are guaranteed a lift home if arrangements fall through. A range of financial incentives also encourages staff to car share. Those joining the scheme receive an M&S voucher worth £20, while those completing six months can choose from a range of car-related perks – the cost of road tax (to the value of the lowest UK charge band) or the same amount of money spent on car servicing or petrol vouchers. Those completing 18 months receive Marks & Spencer vouchers worth £50 (DfT, 2002a).

(a) (b)

Figure 4.6 (a) Reserved car share parking spaces at Boots Beeston site, Nottinghamshire. (b) the car sharer's special parking permit

Potential car sharers at Buckinghamshire County Council can find matches through a centrally coordinated scheme. Four prize draws a year encourage participation, while funds are also set aside for a guaranteed taxi ride home should lift arrangements fall through. Car sharers are exempt from parking charges and can use a 'green bay' space in a nearby car park. Similarly, the Derby-based internet bank, Egg, exempts car sharers from a 75p per day parking charge; in 2003 the proportion of staff sharing cars was about 25%. Publicity for the Buckinghamshire car share scheme emphasises financial savings; for example, one group of sharers was able to use the money they saved to go on holiday (DfT, 2002a).

Car sharing is also promoted at Agilent Technologies just outside Edinburgh, where cars carrying three or more people are able to use dedicated 'green bay' parking spaces located in prime areas. Car sharers initially found matches on a noticeboard but the service is now available on the company intranet. Usage has nearly doubled in five years and regular users from further afield (for example Glasgow) claimed that they saved around £100 a month (DfT, 2002a).

4.6 Travel plans: future directions and implications for policy

In drawing together the key strands of this chapter, it is clear that the existing reach of the travel plan has gradually grown in the UK since its emergence. Indeed, if one attempts to plot how these steps have occurred (see Figure 4.7) it would seem that these stepping stones actually seem to lead towards a possible future policy destination, whereby travel plans continue to develop until:

1. they cover all segments
2. they apply to all proposed and existing organisations (the logical extension from covering all proposed developments as they will in London from 2007, and from them being mandatory for all NHS and government department buildings)
3. they apply to increasingly comprehensive local networks or groups that apply across all segments on a neighbourhood basis.

In other words there is a strong likelihood that travel plans might switch from occupying a small niche, to being not just a mainstream mechanism of transportation demand management, but to being *the* primary means of delivering transport policy within a local area or neighbourhood.

In terms of future implications for policy, such an adjustment to this neighbourhood development approach may finally allow Government to deliver its much publicised sustainable transport policy agenda in a more joined-up and integrated way – rather than in the age-old, mode-by-mode approach which is still very much in vogue.

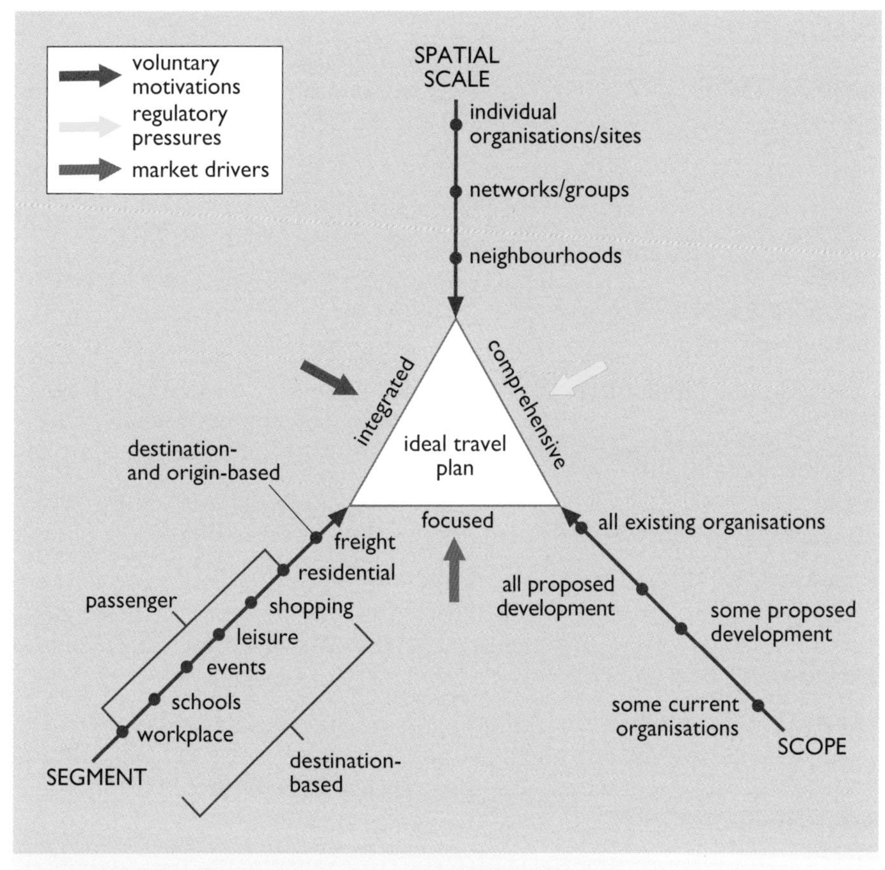

Figure 4.7 Mapping the development of travel plans in the UK

For instance, in London the Local Implementation Plans (LIPs) (equivalent to Local Transport Plans outside the capital) are currently made up of sections considering walking, cycling, parking, etc. and travel plans independently of each other. Instead, LIPs should probably seek to consider local transport issues as a whole on a neighbourhood-by-neighbourhood basis (involving local stakeholders perhaps from some kind of local transport network), look at the authority-wide strategic decisions, and then consider the interactions before finalising the details.

All in all, though, as yet the humble travel plan remains a rather neglected policy instrument that must act in isolation on far too many occasions. One suspects that only with its adoption as the key delivery mechanism for transport policy will it finally live up to its potential as a mainstream mechanism in its own right.

References

Aberdeen City Council (2004) Dyce Transportation Management Association (TMA), Briefing Note, Aberdeen City Council, Aberdeen, 5 August.

Bradshaw, R. (1997) *Employers' Views on Staff Travel Issues, Mobility Strategy Applications in the Community*, University of Westminster, London Transport Studies Group.

Cairns, S., Sloman, L., Newson, C., Anable, J., Kirkbride, A. and Goodwin, P. (2004) *Smarter Choices – Changing the Way We Travel*, London, The Stationery Office.

Caswell, T. (2005) Interview, Associate Director, Vipre, London, 21 January.

Cheshire County Council Travelwise (2002) *Commuter Plans in Cheshire: Steps to success*, Travelwise Team, Cheshire County Council, http://www.cheshire.gov.uk/travelwise/home.htm [no longer available].

Cheshire County Council Travelwise (2006) *Commuter Plans in Cheshire: Steps to success*, Travelwise Team, Cheshire County Council; also available online at http://www.cheshire.gov.uk/Travelplans/stepstosuccess.htm [Accessed 18 September 2006].

Coleman, C. (2000) 'Green commuter plans and the small employer: an investigation into the attitudes and policy of the small employer towards staff travel and green commuter plans', *Transport Policy*, vol. 7, no. 2, pp. 139–48.

Collins, A. (undated) 'Stepping Hill NHS Trust Sustainable Transport Plan', Powerpoint presentation, Stepping Hill NHS Trust.

Department of the Environment, Transport and the Regions (DETR) (1998) *A New Deal for Transport: Better for Everyone*, London, The Stationery Office.

Department of the Environment, Transport and the Regions (DETR) (1999) *Preparing your Organisation for the Future: The Benefits of Green Transport Plans*, London, The Stationery Office.

Department for Transport (DfT) (2002a) *Making Travel Plans Work: Case Study Summaries*, London, The Stationery Office; also available online at http://www.dft.gov.uk/stellent/groups/dft_control/documents/contentservertemplate/dft_index.hcst?n=13896&l=4 [Accessed 29 August 2006].

Department for Transport (DfT) (2002b) *Vodafone*, [online], London, DfT. http://www.dft.gov.uk/stellent/groups/dft_control/documents/contentservertemplate/dft_index.hcst?n=13896&l=4 [Accessed 18 September 2005].

Department for Transport (DfT) (2003) *Expert Advice Helps Cut Fleet Costs, Transport Energy Best Practice, Good Practice Case Study 409* [online], London, The Stationery Office; http://www.rmd.dft.gov.uk/project.asp?intProjectID=11060 [Accessed 18 September 2006].

Department for Transport (DfT) (2005) *Making Residential Travel Plans Work: Good Practice Guidelines for New Development*, London, DfT.

Department for Transport, Local Government and the Regions (DTLR) (2001a) *National Travel Survey 1998–1999*, London, The Stationery Office.

Department for Transport, Local Government and the Regions (DTLR) (2001b) *Take Up and Effectiveness of Travel Plans and Travel Awareness Campaigns*, London, DTLR.

Energy Efficiency Best Practice Programme (2002) *A Travel Plan Resource Pack for Employers*, Energy Efficiency Best Practice Programme, London, The Stationery Office.

Enoch, M.P., Zhang, L. and Morris, D. (2005) *Organisational Structures for Implementing Travel Plans: A Review, Research Report for OPTIMUM2 Project, London Borough of Southwark Cluster*, European Regional Development Fund Interreg IIIB Programme, Lille, May; also available online at http://www.optimum2.org [Accessed 4 September 2006].

Ginger, M. (2005) Telephone interview, Bristol City Council, Bristol, 4 March.

Martz, J. (2006) 'The American Experience', Paper presented to the *Association of Commuter Transport Conference*, London, 21 February.

Morris, D. (2005) *Care Free Development: The Potential for Community Travel Plans*, Universities Transport Studies Group Conference, University of West of England, Bristol, 5–7 January.

Murphy, W. (2005) Personal communications, Aberdeen City Council, Aberdeen, 27 January and 22 March.

NHS Estates (2001) *Sustainable Development in the NHS*, London, The Stationery Office.

North East Scotland Transport Partnership (NESTRANS) (2004) *NESTRANS Appoint Travel Awareness Project Leader* [online], NESTRANS, News Release, Aberdeen, 5 July, http://www.nestrans.org.uk. [Accessed 4 September 2006].

Office of the Deputy Prime Minister (ODPM) (2005) *Planning Obligations*, Planning Circular 05/05, London, Office of the Deputy Prime Minister, 18 July. http://www.communities.gov.uk/index.asp?id=1500145 [Accessed 18 September 2006].

Rye, T. (2002) 'Travel plans: do they work?', *Transport Policy*, vol. 9, no. 4, October, pp. 287–98.

Sustrans (2006) *Safe Routes to School: Case Studies* [online], http://www.saferoutestoschools.org.uk [Accessed 14 July 2006].

Transport 2000 (1998) *The Healthy Transport Toolkit: A Guide to Reducing Car Trips to NHS Facilities*, London, Transport 2000 Trust.

Transport 2000 (2001) *Tourism Without Traffic: A Good Practice Guide*, London, Transport 2000 Trust, September.

Transit (2002) 'First offers Travel Plans to Hants businesses', *Transit*, no. 191, 6 September, p. 32.

Conclusion

by Stephen Potter and James Warren

Conclusion – from here to 'ecoternity'

Backcasting from sustainability

The broad plan of this book has been to start with a 'backcasting' exercise to identify what sort of overall approach is needed to meet transport's energy challenge. Getting from here to an 'ecoternity' of sustainable transport requires a combination of technical, organisational and behavioural changes. This book has explored the role of these key factors in achieving a sustainable future. The simple backcasting model in Chapter 1 concludes that it is necessary to combine various technical and behavioural change approaches to have any hope of achieving a sustainable transport system. Other backcasting scenario studies (e.g. Hickman and Banister, 2006 and 'Visions for the future' in Banister, 2005) have come to the same conclusion.

Subsequent chapters have examined in more detail the role of technical developments and behavioural change. Chapter 2 noted that fuel formulations and cleaner vehicle technologies have substantially reduced pollutants from vehicles, producing significant improvements to air quality in towns and cities. Some persistent air quality problems remain, particularly as vehicle numbers grow, but the most substantial challenge is to significantly cut CO_2 emissions from road transport. Trends towards more powerful cars have counterbalanced improvements in engine designs (Cousins et al., 2006). In the USA, sport utility vehicles (SUVs) have become something of a cult object, gaining huge market shares in recent years. SUVs have a fuel consumption nearly a third worse than passenger cars. In 1975 SUVs represented less than 2% of all light vehicle sales. By 2005 this had increased to over 25% of the market (Davis and Diegel, 2006).

Although less extreme, European car purchasing trends have also concentrated on higher performance rather than fuel economy. Consequently car purchasing trends are set to fail the 2008 and 2012 EU targets to cut test CO_2 emissions from new cars. The EU policy for a 33% improvement is technically possible, but people are simply not buying low-carbon and fuel-efficient cars.

This provides a political dilemma. Should the consumer's love of the power, size and performance of cars be taken as a given, or is it necessary for motorists to accept a different sort of car? In the latter case behavioural change is needed and this is where technical approaches are facing strong resistance. It is notable that the most successful lower-carbon cars and technologies are ones that do not challenge the motoring regime of power, size and performance. Hybrid cars offer both power and a potential 30% improvement in fuel economy (in practice probably somewhat lower (E4tech, 2006)). Such a compromise is also offered by some alternative fuels. LPG and bio-diesel are beginning to establish niche markets by providing power with reduced CO_2 emissions. But the big question is how far this compromise with the current type of car takes us in terms of reducing CO_2 emissions. In the next 20 years it seems that a more diverse mix of fuels will emerge, but lower-carbon alternative fuels seem unlikely to be more than 20% of transport fuel consumed (*Fuelling Road Transport*, Eyre et al., 2002). This would cut the entire car fleet's CO_2 emissions by only about 6%.

Technologies that require behavioural change to accept a different type of car have had little impact. Battery electric vehicles are unlikely to have a long-term future; not only are they lower powered and performing than petrol and diesel cars, but they have range limitations and high capital costs. Fuel cells are now set to emerge as the main challenger to the internal combustion-engined car. However, the role of hydrogen fuel cell vehicles remains the great unknown of transport futures. Their ability to reduce CO_2 emissions depends crucially on how hydrogen is produced; they will also be expensive to buy and their fuel supply infrastructure could be problematic. Even given the 'rapid progress' scenario from *Fuelling Road Transport*, with fuel cell vehicles taking a 20% share of the 2020 car market, the net effect would, at best, be an overall cut of 12% in the car fleet's CO_2 emissions. Adding the above effects together suggests that all the alternative fuel technologies might, given strong support and political will, approach the target specified in the index model in Chapter 1 for a 20% cut in carbon intensity per vehicle kilometre over a 20-year period.

But the index model showed that low carbon fuels need to be combined with a large improvement in fuel economy to achieve a sustainable transport system. A major improvement to fuel economy is technically possible, but requires an acceptance by motorists of a change in the size and/or performance of cars. This brings us back to the issue of technical approaches that also require behavioural change from the motorist – particularly an acceptance of lower-powered and lower-performing cars. Some technologies, such as hybrids, can maintain performance and power by improving fuel efficiency. However, this can be taken only so far. Ultimately, transport sustainability challenges our ingrained obsession with a car's power, acceleration and performance. Ultimately, technical approaches cannot succeed without a change in attitudes and behaviour towards our view of the car. To date, policymakers have ducked the issue and favour the less politically contentious option of fuel switching. Thus the current path we are embarking upon is for petrol and diesel 'gas guzzlers' to be replaced by alternative fuel guzzlers.

Current trajectories

Current trends and the various technical and behavioural change options explored through the formula model detailed in Chapter 1 can be represented as paths on a backcasting diagram. Figure c.1 (based on an illustrative diagram in Hickman and Banister, 2006) shows the general approach diagrammatically. 'A' is where we are now, with the curve up to this point representing the growth in transport's environmental impacts to date. We are already well above the sustainability zone. The top curve represents current 'business as usual' trends, whereas the lower curve is the sort of path needed to return to a sustainable level of environmental impacts. This has year 'C' set for achieving a sustainability target, with an interim target by year 'B'.

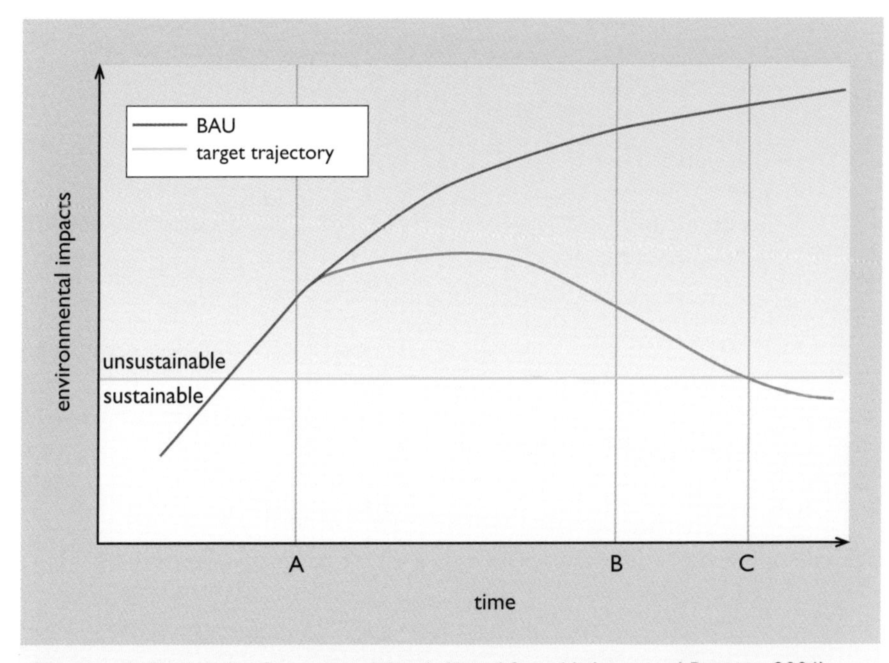

Figure c.1 Backcasting for sustainability (adapted from Hickman and Banister, 2006)

The formula model in Chapter 1 started by noting current trends and projections. Figure c.2 maps onto the backcasting graph the UK's actual transport CO_2 emissions and three projections for future trends. The sustainability zone is the target identified in Chapter 1, based on the various IPCC reports (summarised in Houghton, 2004). Hickman and Banister (2006) in their backcasting exercise use a 60% reduction target based on the VIBAT (Visioning and Backcasting) research for the Department for Transport's 'Horizons' research programme (UCL, 2006).

The curve to date is actual CO_2 emissions from transport in the UK and the top line is the government's 'business as usual' projection (from *Transport Statistics Great Britain*, DfT, 2004a). The bottom projection is the anticipated effect of policies included in the 2004 Transport White Paper (DfT, 2004b). It is notable that this projection would only return transport CO_2 emissions to their 1990 levels by 2030. Already one key element of meeting this projection, the EC motor industry voluntary agreement to improve fuel economy and CO_2 emissions, has failed. It is notable that the 2006 Energy Review (published as *The Energy Challenge*) provides a less optimistic projection and anticipates emissions from transport continuing to grow to 2015 and thereafter to fall (DTI, 2006, p. 126). This is the middle 'fuel switch' projection. The Energy Review looks to an unspecified successor to the voluntary agreement and more significant reductions after 2020 from more advanced technological developments.

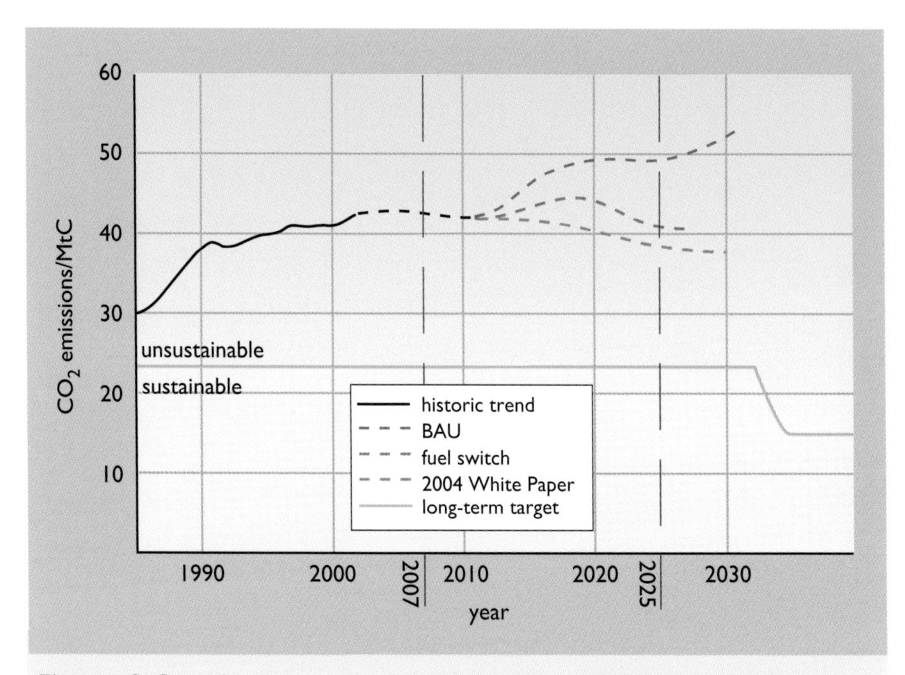

Figure c.2 Ground transport carbon dioxide emissions 1970–2004 and selected projections (source: DEFRA, 2004, Figure 7; DfT, 2004b; DTI, 2006)

Both the over-optimistic 2004 Transport White Paper projection and that of the 2006 Energy Review represent an improvement on 'business as usual'. Indeed, the latter notes that without the policy actions taken, CO_2 emissions from transport would have been 15% higher than projected for 2010. However, Figure c.2 also includes the sustainability target used in this book, derived from the IPCC scientific assessment of the reduction in CO_2 emissions needed to moderate the impacts of climate change. This is for an initial cut of 40% from 1990 levels and eventually a cut of 60%. The sort of policies currently being pursued may, at the very best, return transport's CO_2 emissions to near 1990 levels, but they fall hopelessly short of even the short-term sustainability target.

This indicates the limitations of policies concentrating on technologies that do not challenge the current car culture of power, acceleration and performance. These projections are for a future of 'low carbon fuel guzzlers', and that does not take us 'from here to ecoternity'. Emissions arising from increases in the volume of travel more than counterbalance the shift to low-carbon fuels. It may be politically astute to concentrate upon policies for technologies that do not require a change in motorists' attitudes and behaviour, but these alone will not get anywhere close to a sustainable transport system.

So, as was concluded in Chapter 2 of this book, fuel switching to reduce carbon emissions by 20% in 20 years looks viable, and policies and industry responses are coming into place for this. The second area of technical improvements, a substantial improvement in fuel economy, is a totally different matter. This is technically viable, but to succeed requires a change in motorists' attitudes and behaviour. Thus the issue of behavioural change is one that is as crucial to the success of technical solutions as it is for mobility management approaches.

Alternative trajectories

The final scenario using the formula model suggested that, as well as a 20% reduction in carbon intensity through fuel switching, something like a 50% improvement in car fuel economy should be sought (plus a 40% improvement to bus and train fuel economy). If this were done then, over 20 years, it would get us significantly towards a sustainable level of transport CO_2 emissions, at least for the short term. Figure c.3 maps such a combined fuel switch and fuel economy trajectory onto the backcasting graph.

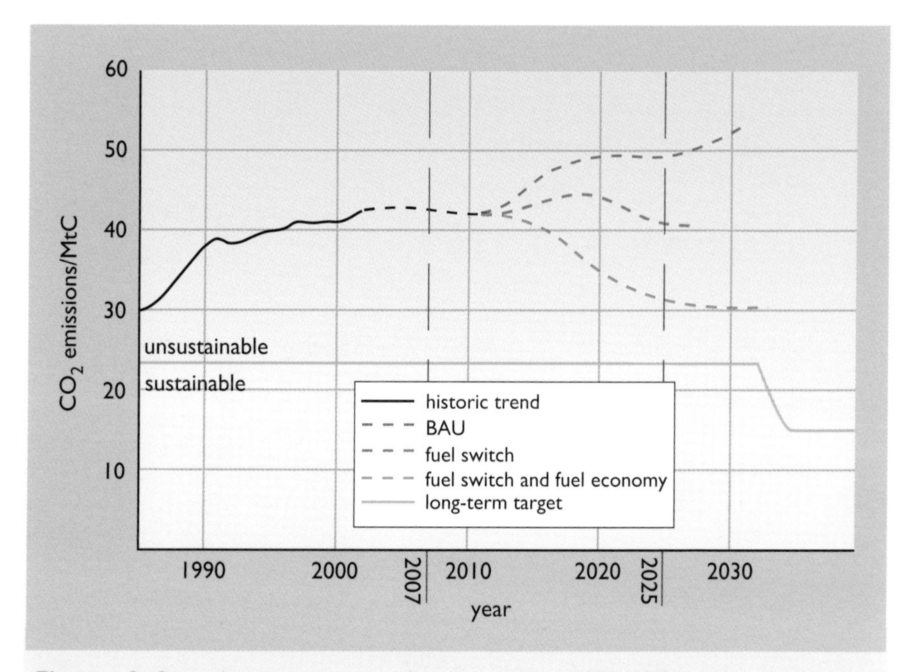

Figure c.3 Ground transport carbon dioxide emissions 1970–2004 and key projections (source: DEFRA, 2004, Figure 7; DTI, 2006)

The issue of behavioural change to realise the benefits of low-carbon and fuel-efficient technologies is clearly vital, and of course is central once mobility management policies are considered. Mobility management seeks to influence a range of factors that together generate demands for travel. This involves a complex set of economic and social factors that interact with each other at a number of levels. Figure c.4 is a representation of this interrelated system and some of the links involved. The diagram has groups of 'causes' towards the outside (coloured brown) and transport results/effects in the centre (coloured white). These spill out in terms of travel volume and CO_2 emissions (coloured purple). This is an illustrative simplified diagram, but even so it expresses some of the issues and complexity involved when policy measures seek to intervene in such a system. Policies need to target the brown boxes, with success indicated by the impacts upon those coloured white and purple.

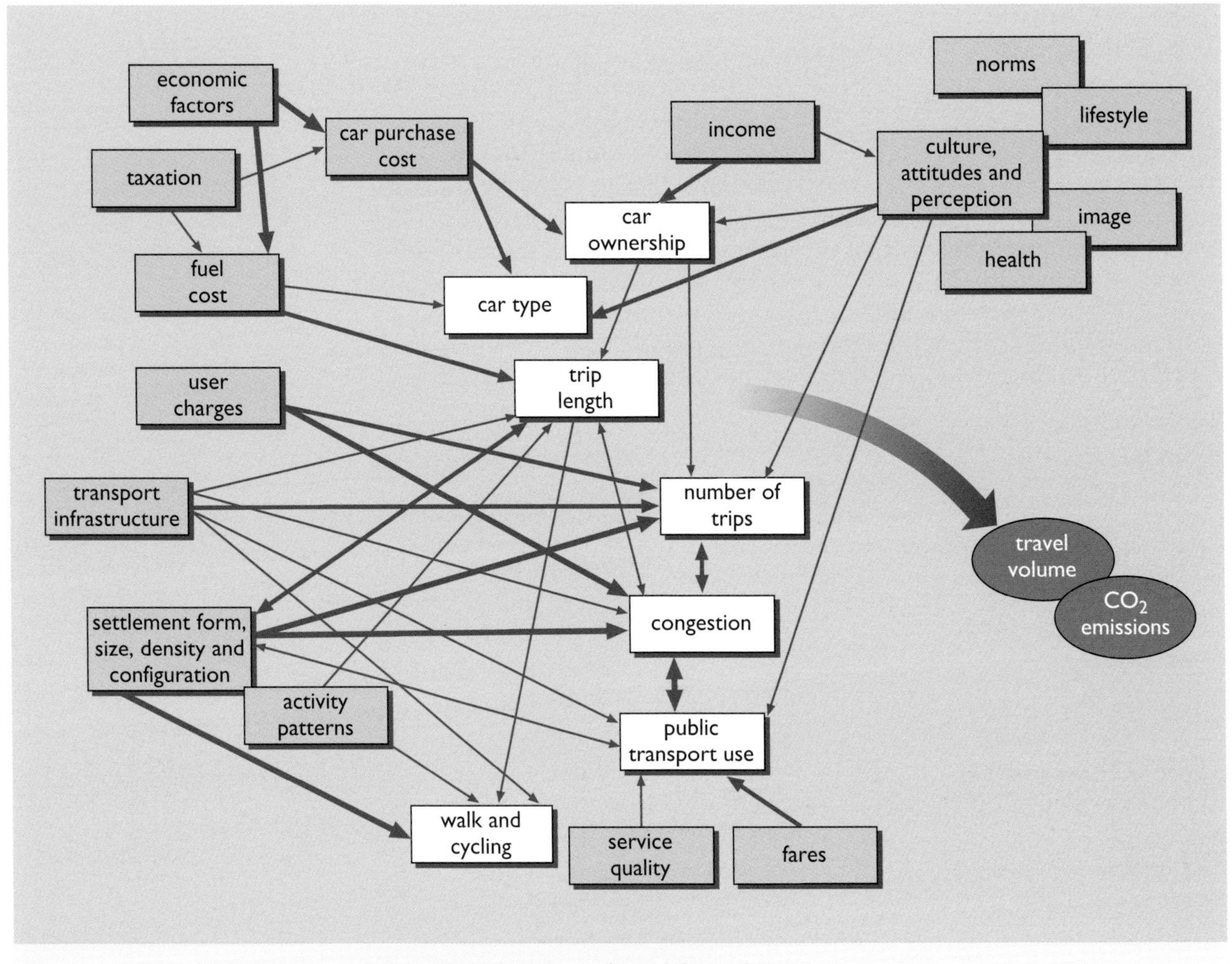

Figure c.4 Factors contributing to the complex interactions of travel demand

Many factors both influence travel and are influenced by it (e.g. settlement form, size and density is both a determinant of travel patterns and, over time, itself is determined by changes in travel behaviour). Some travel-determining factors, like income and economic factors, are more independent and some are more amenable to policy influence than others. For example, central and local government have strong control over the provision of transport infrastructure (roads, metro lines or cycle paths) but only a weaker influence over settlement patterns through planning and development control functions. Transport taxation can affect part, but not all, of the costs of travelling. Here, targeted incentives can be important, such as tax breaks to help consumers purchase new technologies such as hybrids or fuel cell vehicles.

Overall, the potential for influence varies considerably and, because this system is extremely dynamic with numerous feedback effects, only concerted, linked action across several factors is likely to be effective. In practice actions are often contradictory and counterbalance each other. For example, there may be investment in rail and tram infrastructure together with better bus services to promote modal shift from the car, but then fares are increased to pay for this and motoring costs are reduced by a cut in tax on greener fuels. All these separate actions work in contradictory directions.

Mobility management approaches have been developed to address several of the factors in Figure c.4. To the top left of the diagram are economic factors, and mobility management measures here might include taxes on fuel, on vehicle purchase and ownership. These could be designed in particular to influence vehicle choice (e.g. tax concessions or subsidies on low-carbon vehicles) or be more general to try to influence modal choice or the volume/length of trips (e.g. a general fuel tax). User charges, such as for parking, the London Congestion Charge or road and bridge tolls, tend to be more specific and focus mainly upon trip numbers and modal choice rather than vehicle type.

To the bottom left are factors around the way in which settlement patterns, density and transport infrastructure affect the amount of car and public transport use (and walking and cycling as well), and also the level of congestion. At the top right of the diagram are cultural attitudes and perceptions, which are moulded and influenced by a whole host of factors, only some of which are mentioned here. Informational measures may be used here, for example linking travel behaviour to health and children's well-being.

Behavioural change approaches

The behavioural change approaches covered in Chapters 3 and 4 concentrate upon one particular type of mobility management. This is the role that big generators of travel demand, such as employers, service providers, schools and universities, can have in managing and reducing the environmental impact of travel to their sites. Such institutional mobility management (travel plans) provide a good example of trying to intervene across key parts of the travel generation system. However they, of course, represent only one of a number of behavioural change responses and policies. As they are implemented by institutions, travel plans affect only certain trips (particularly commuting and travel to major services such as hospitals) and other policies would need to cover other trip types and have a general impact across the system as a whole (which would include, for example, transport taxation, investment in energy-efficient and low-carbon transport modes, planning policies that reduce travel needs, etc.).

However, although they are but part of the new transport policy mix, travel plans illustrate well the challenges and issues involved in developing 'new' transport policy approaches. In particular they include addressing the difficult issues in the top right corner of Figure c.4, concerning attitudes and perceptions to travel. Travel plans are part of a shifting from a 'top down' approach of policymakers deciding what is needed for people and institutions, to where people and institutions play a much larger role in 'owning' the transport problem and actively being part of its solution.

The impacts of travel plans are mixed, largely because they vary immensely in quality and the seriousness with which the institutions implement them. If travel plans are well managed, they can be an effective tool for reducing car use. A study that pulled together evidence on the impact of travel plans (Rye, 2002) found that travel plan effectiveness varied depending on the measures that were implemented:

- a plan containing only marketing and promotion was unlikely to achieve any modal shift
- a plan with car-sharing measures may achieve 3–5% reduction in drive-alone car commuting
- a plan with car sharing, cycling and large discounts (more than 30%) on public transport plus works buses will achieve a 10% reduction in drive-alone car commuting
- the combination of the above measures, together with disincentives to drive, can achieve a 15–30% reduction in drive-alone commuting.

In practice, most travel plans do not progress beyond the less-effective levels. They have not been widely adopted, and they are introduced with some reluctance. Rye (2002) estimated that travel plans removed about 150 000 car trips each working day from British roads, or 1.14 billion kilometres per year. This is not a lot: it equates to under 1% of the total vehicle journeys to work. Rye goes on to suggest that this low take-up to date is due to five major factors:

- companies' self-interest and internal organisational barriers
- lack of regulatory requirements for travel plans
- personal taxation and commuting issues
- the poor quality of alternatives in the UK (particularly public transport)
- lack of experience due to the novelty of the concept.

This returns us to the point made when looking at Figure c.4 that it is necessary to have policies acting together across the travel generating system and that often progress can be blocked by counteracting factors elsewhere in the system.

There is potential in the travel plan concept. If travel plans are well designed and implemented consistently, they can reduce single-driver car use by at least 10–20% and possibly more. The target for modal shift in the final (sustainability) version of the formula model is 23%, so the best travel plans are approaching what is needed from their sector (although a similar success rate would be required from other mobility management measures as well).

However, such successful travel plans are rare. There is a need to extend good practice and to integrate travel plans with other behavioural change measures, such as general investment in public transport, tax changes to reward 'green' travel, and measures to cut the distances we need to travel. In addition, of course, more general travel behaviour measures that address individuals are necessary, in order to hit the target in the Chapter 1 index model. Nevertheless, it is clear from the evidence here that achieving such a target is possible. Whether institutions, government and individuals consider the environmental and congestion costs of our transport problems to be sufficiently serious to merit such actions is another matter.

Figure c.5 completes the backcasting exercise 'from here to ecoternity'. A combined strategy of complementary policies and actions is needed, with roughly equal contributions coming from fuel switching, fuel economy improvements and modal shift/trip reduction.

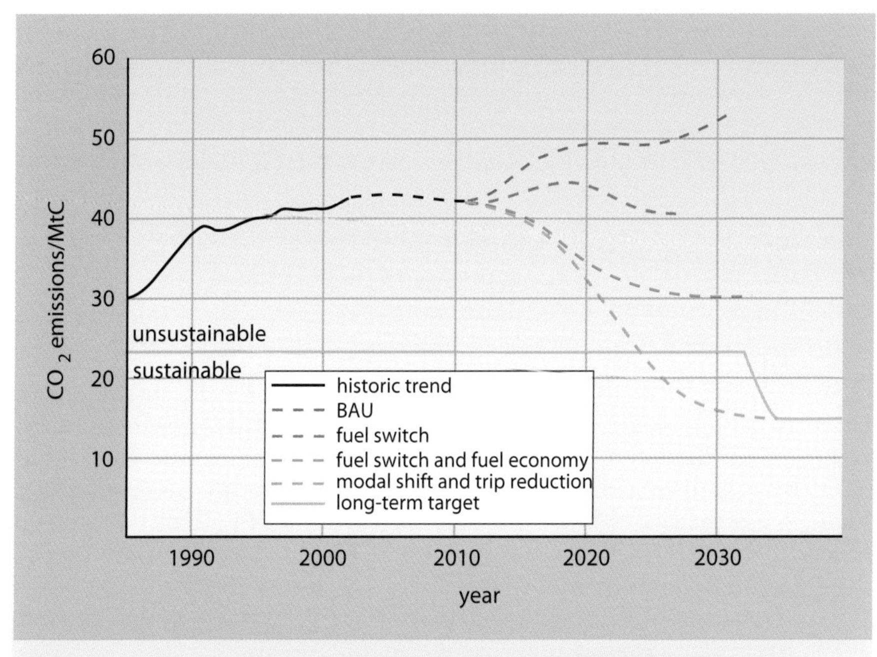

Figure c.5 Ground transport carbon dioxide emissions 1970–2004 and possible 'sustainable' projections (source: DEFRA, 2004, Figure 7; DTI, 2006)

This is all well and good, but in reality we are nowhere near the projection that leads to sustainability. As was noted at the beginning of this chapter, at best we are running along the 'Fuel Switch' projection, which itself is a retreat from the more optimistic projection in the 2004 Transport White Paper. Measures to address behavioural change, that are needed both to improve fuel economy and for modal shift/trip reduction, are weak, widely resisted and used effectively in only a tiny proportion of situations. When used well, they work, as measures like the London Congestion Charge, some travel plans and a few city integrated transport/planning schemes demonstrate (Banister, 2005). But these are only isolated niches, not widespread practice.

Drivers for transformation

Environmental drivers

We are not as yet on a path to sustainable transport, but there are forces emerging that might drive us that way. Firstly there is the growing political awareness of the impacts of climate change. It may now be too late to stop some degree of global warming, but computer projections are showing the odds shortening on frightening worst-case scenarios (Houghton, 2004). These include the desertification of large parts of the USA and even the Amazon, the loss of major coastal cities and the prospect of mass international migration to developed countries. Avoiding such worst-case scenarios is still possible with radical action.

This growing environmental awareness seems to be behind some individual states in the USA committing themselves to a major reduction in CO_2 emissions (notably California's commitment in 2006 for up to an 80% cut). A rather fascinating indicator of the shift in long-entrenched attitudes comes from the high-performance world of Formula One motor racing.

The regulations for engine design are to be changed to require energy efficiency and energy recovery (BBC Sport, 2006). For a long time motor racing has been the test bed for new car technologies, with the design specification for engines being the development of power, acceleration and high performance. Redirecting research towards an agenda of energy efficiency marks a significant shift.

Health and congestion

But it probably will not be the environmental agenda alone that will kick off a transformation in our use of transport. Behind many of the above seemingly 'green' transport initiatives are more powerful drivers.

This can be illustrated by the example of travel plans that were particularly examined in Chapter 4 of this book. School travel plans have been taken up with markedly more enthusiasm than those for workplaces (Emmerson, 2006). A key reason for this is that school travel plans link into very major concerns for children's health and the alarming rise of obesity among children. They have thus been seriously supported by Local Education Authorities and the schools themselves. Workplace travel plans have not linked to similar strategic concerns within businesses.

At a more general level, the economic impacts of traffic congestion have ended up being the biggest influence on the design of mobility management measures. Traffic congestion has for long been an important economic issue and from the 1930s has been the core rationale for improving, widening and building new roads. At about the same time as transport's environmental concerns emerged it was realised that road building was failing to reduce congestion and support economic growth. A seminal piece of research was the 1991 report *Transport: the New Realism* (Goodwin et al., 1991). This showed that Britain could not physically or economically accommodate the then Department of Transport's road traffic forecasts for a 142% increase in traffic to 2025. This conclusion was supported by several government reports and in a later report, Goodwin (1994) used research by the pro-road building British Road Federation to show that even a massive road construction programme (beyond that which Britain's economy could sustain) would fail to stop congestion getting worse. At about the same time the Confederation of British Industry (CBI) estimated traffic congestion to cost the UK economy £15 bn per annum.

In a densely populated country like the UK, mobility management is inevitably needed to control traffic congestion, and this is also the case in many developed and developing economies – including major population centres of China and India.

In the UK, the 1990s saw the government cut road building programmes and increase fuel duty on petrol and diesel. The latter was abandoned following the fuel duty protests in 2000, with no effective policy to replace it. Some road building recommenced, but with the success of the London Congestion Charge, and the failure of less radical policies to make any impact on congestion, the idea of a national road pricing scheme gained sudden acceptance. In July 2004, UK Transport Secretary, Alistair Darling, announced that replacing Road Fuel Duty and Vehicle Excise Duty (VED) with some form of widespread road user charging was envisaged

(DfT, 2004c). Since then, UK plans for national road user congestion charging have been firming up, with plans for in-vehicle instrumentation for all cars within 10 years preceded by a series of regional pilot projects. The question is not of 'if' national road user charging will be introduced, but 'how' and 'when'.

The introduction of national road user charging to replace existing fuel and other car taxation could produce environmental benefits, but this is not the core purpose of the measure. It is about cutting road congestion and this will not necessarily reduce the environmental impacts of transport. The current design of the charging system is to provide a pricing system that will 'even out' traffic flows away from peak periods towards less congested times and places. The charge per mile would vary only by congestion – motorists would pay more for travelling during busy peak hours and on roads and in places most prone to congestion (e.g. in and around large towns and cities). The charge for travelling in rural areas and outside peak hours would be low. Indeed, if the road user charge entirely replaced fuel tax, rural motoring would become cheaper than it is today.

Such a charge based only on congestion would not vary according to the fuel economy of a car or the carbon content of the fuel used. Thus a highly fuel-inefficient 'gas guzzler' would pay the same amount as a low-carbon or fuel-efficient car. With the national congestion charge proposed to replace fuel tax, this would eliminate the automatic incentive that fuel tax provides for fuel-efficient cars. A further weakness is that patterns of activities would alter in response to changes in transport costs. Motorists would shift to driving to destinations in low-charge areas, increasing trip lengths and fuel consumption.

The 2004 Transport White Paper (DfT, 2004b) conceded that the policy for national road user pricing was not to address CO_2 emissions and that there was uncertainty about whether road pricing would increase or decrease emissions. The Paper also noted that the most effective way of reducing CO_2 emissions from transport would be to take measures that affect the cost of fuel and the price of energy-efficient vehicles. Subsequently, proposals are being discussed for complementary measures to support low-carbon and fuel-efficient cars. This could be retaining fuel duties, but politically it would be very hard to introduce a national road user charge without reducing other motoring taxes, of which fuel duty is the largest. This has led the UK to examine the sort of car purchase and ownership taxation measures already used in some EU states (Potter and Parkhurst, 2005). Most EU states have a special car purchase tax in addition to VAT. In some countries this is used to favour fuel-efficient and low-carbon cars. In Finland, for example, there is a reduction in car purchase tax for low-emission vehicles. In the Netherlands, car purchase tax is 45.2%. This may seem high (although at 105% Denmark's is higher), but there are counterbalancing fixed allowances of €1540 for petrol and LPG cars, €580 for diesel cars and other allowances for low-carbon vehicles. This fixed allowance cuts the charge significantly for low-carbon and more fuel-efficient cars whilst having little impact on the price of larger and less fuel-efficient vehicles.

The UK is not at the moment looking to reintroduce car purchase tax (which was abolished in 1992), but it appears that a cross-party consensus is emerging to further reform the CO_2-based VED (or 'car tax'). This could be highly graded, with mention being made of a top charge of £1800 per annum for the least fuel-efficient cars (Cooper, 2006).

Overall, a complex mix of responses is emerging as governments are proposing and making changes to transport taxation. Proposed policies reflect powerful economic and social drivers, with issues of motorist acceptance being very important. The real danger is not that a road user congestion charge that is environmentally degrading could be introduced but that the counterbalancing 'green' purchase/VED measures are rejected or watered down so that they are of little use. Economic issues around congestion are a powerful driver for transport policy, but whether this will produce a serious step towards transport sustainability is far from clear.

Global economic drivers

It seems possible that in the next few years we may see stronger purchase or ownership tax measures to promote fuel switching and possibly also fuel economy. However it will probably not be environmental policies, taxation, actions of industry or key institutions that will be the strongest drivers towards sustainable transport. Other, more powerful factors (like congestion reduction) will spur government, institutions and individuals into serious action. But there are also significant global economic trends that are going to have major impacts on our travel systems and behaviour. The first and most obvious of these is the increase in oil prices. The price of crude oil tripled from $25 a barrel in 2000 to $75 a barrel in 2006 and the economic forces behind this rise are set to continue. Rising oil prices have now replaced rising fuel duties as the main factor increasing the cost of petrol and diesel. Key to the oil price rise is the burgeoning growth of China, India and other Asian economies. This price rise has occurred when oil supply has broadly been able to meet demand, albeit with some significant political glitches. However this is unlikely to be the case for very much longer. As was mentioned in the Introduction, we are approaching the peak in global oil production (Figure c.6), which is predicted to occur some time between 2008 and 2015 (Laherrère, 2001; BP, 2003; Crabbe, 2003). It is salient to compare Figure c.6 with that of the demand for oil (Figure i.1) in the Introduction to this book. Just as the new economies of the East take off, oil production will stagnate and then start to drop. An inevitable consequence is that energy costs are set to rise dramatically.

In the new economic order of energy supplies unable to match demand, there can be two possible responses. One is for nations to secure their own supplies by whatever means possible. It seems that the nations of the West are already embarking upon this approach; witness the emergence in the last few years of 'energy security' onto the international political agenda. The alternative is that energy efficiency will become central to economic development. As part of the economic rise of the East, the demand for motorised transport is also taking off, but it may well prove impossible to meet this demand using the current transport model and system. While the West seeks to secure oil supplies and obtain new energy sources to carry on in the old ways, it may well be that the economies of the East will adopt a different approach. Less able to muscle in upon dwindling oil reserves, they may be the ones to reinvent transport because the old way of

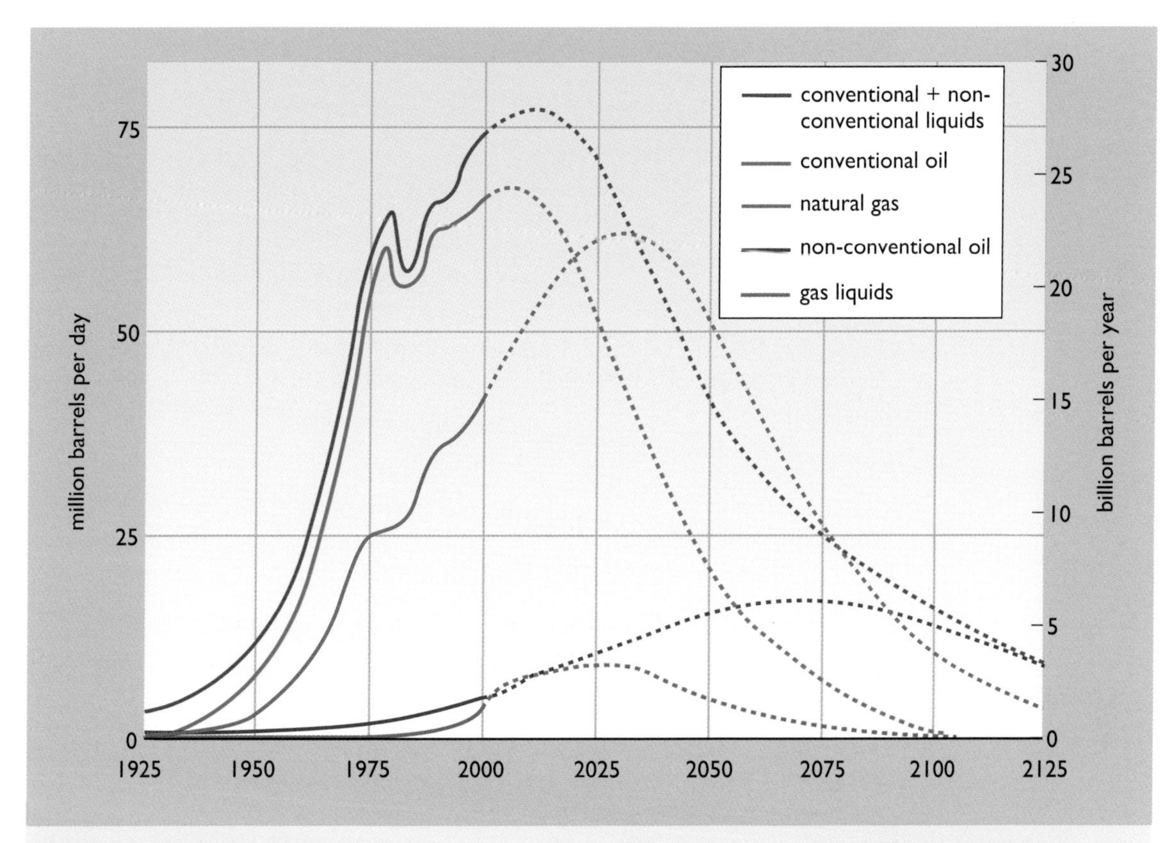

Figure c.6 Global production of oil and gas, after Laherrère, 2001. Figures to the year 2000 are historical data; thereafter the dotted curves represent projections of future supply.

doing things will be too costly and inefficient. The East has already outcompeted the West in transport by doing things in new and innovative ways (witness the rise of Japanese, Korean and now Chinese car industries). Reinventing the transport system may become yet another way in which the East will outcompete the West. And, just in passing, one result could be transport moving closer to being a sustainable activity.

References

Banister, D. (2005) *Unsustainable Transport*, Oxford, Routledge.

BBC Sport (2006) '*Green' engines given F1 go-ahead* [online], http://news.bbc.co.uk/go/pr/fr/-/sport1/hi/motorsport/formula_one/5253748.stm [Accessed 8 August 2006].

BP (2003) *BP Statistical Review of World Energy* [online], BP, http://www.bp.com/centres/energy2002/index.asp [Accessed 22 May 2006].

Boyle, G., Everett, R. and Ramage, J. (eds) (2003) *Energy Systems and Sustainability*, Oxford, Oxford University Press/Milton Keynes, The Open University.

Cooper, K. (2006) 'MPs call for £1,500 car tax', *The Times*, 7 August.

Cousins, S.H., Garcia Bueno, J. and Palomares Coronado, O. (2006) 'Powering or De-Powering future vehicles to reach low carbon outcomes: the long term view 1930–2020', *Journal of Cleaner Production (in press).*

Crabbe, R. (2003) 'Oil and Gas', in Boyle et al. (eds) (2003).

Davis, S. C. and Diegel, S. W. (2006) *Transportation Energy Data Book*, US Dept of Energy ORNL, USA.

Department for Environment, Food and Rural Affairs (DEFRA) (2004) *e-Digest of Environmental Statistics* [online], http://www.defra.gov.uk/environment/statistics/globatmos/download/xls/gafg07.xls [Accessed 12 August 2006].

Department for Trade and Industry (DTI) (2006) *The Energy Challenge*, London, DTI.

Department for Transport (DfT) (2004a) *Transport Statistics Great Britain*, London, Department for Transport.

Department for Transport (DfT) (2004b) *White Paper: The Future of Transport: a Network for 2030*, London, Cmd 6234, DfT.

Department for Transport (DfT) (2004c) *Feasibility Study of Road Pricing in the UK: A Report to the Secretary of State*, DfT.

Emmerson, G. (2006) 'Back to school for workplace travel plans?', *Local Transport Today*, 20 April, p. 15.

Eyre, N., Fergusson, M. and Mills, R. (2002) *Fuelling Road Transport: Implications for Energy Policy*, London, Energy Savings Trust.

E4tech (2006) *UK Carbon Reduction Potential from Technologies in the Transport Sector, Final Report for the UK Department for Transport.* E4tech, May.

Goodwin, P.B. (1994) *Traffic Growth and the Dynamics of Sustainable Transport Policies*, Transport Studies Unit, Oxford, Oxford University.

Goodwin, P., Hallett S., Kenny, F. and Stokes, G. (1991) *Transport: the New Realism*, Transport Studies Unit, Oxford, Oxford University.

Hickman, R. and Banister, D. (2006) 'Looking over the horizon', Town and Country Planning, vol. 75, no. 5, May, pp. 150–2.

Houghton, J. (2004) *Global Warming: the Complete Briefing,* Cambridge, Cambridge University Press.

Laherrère, J. H. (2001) 'Forecasting future production for past discovery'. *OPEC seminar*, 28 September.

Potter, S. and Parkhurst, G. (2005) 'Transport policy and transport tax reform', *Public Money and Management*, vol. 25, no. 3, June, pp. 171–8.

Rye, T. (2002) 'Travel plans: do they work?', *Transport Policy*, vol. 9, no. 4, October, pp. 287–98.

UCL (2006): *Visioning and backcasting for UK transport policy* [online], http://www.bartlett.ucl.ac.uk/research/planning/vibat [Accessed 30 August 2006].

Acknowledgements

Grateful acknowledgement is made to the following sources:

Chapter 1

Tables

Table 1.9: Potter, S. (2003) 'Transport Energy and Emissions: Urban Public Transport', in Hensher, D. and Button, K. (eds.) *Handbook in Transport 4: Transport and the Environment*, Pergamon Press, Elsevier Science Ltd.

Figures

Figures 1.1, 1.4 and 1.7: © Stephen Potter; Figure 1.3: © Press Association/ Barry Batchelor; Figure 1.5: Courtesy of John Hopkins University Applied Physics Laboratory.

Chapter 2

Text

Box 2.4: DEFRA (2002) 'Air pollution – what it means for your health'. Crown copyright material is reproduced under Class Licence Number C01W0000065 with the permission of the Controller of HMSO and the Queen's Printer for Scotland; Pages 57–58: Adapted from Zakian, M. (2006) 'Gentlemen charge your engines,' *The Guardian*, 15th June. © Guardian Newspapers Limited 2006; Box 2.9: Adapted from *Case Study: LPG Vehicles: Oxfordshire Mental Healthcare NHS Trust*, The Energy Saving Trust; Box 2.10: Wainwright M. (2002) 'Chicken Fat to Power Lorries', *The Guardian*. © Guardian News and Media Limited 2002; Box 2.13: Adapted from McKay, A. 'The G-WIZ automatic electric vehicle', *Scotland on Sunday*, 30 April 2004. © Alastair McKay.

Tables

Table 2.2: DieselNet (2006) *Emissions Standards (international)*, www.dieselnet.com/standards

Figures

Figure 2.1: © National Motoring Museum; Figure 2.2: © David Noble Photography; Figure 2.5: © Reed Hecht; Figure 2.11: Courtesy of Toyota (GD) plc.; Figure 2.12: Courtesy of Low CVP; Figure 2.13: Courtesy of David Lewry, Cheshire County Council; Figure 2.14: © Ecoscene/Joel Creed; Figure 2.15: © Graham Jepson; Figure 2.16: © Martyn Goddard. Courtesy of GoinGreen; Figure 2.17: Oak Ridge National Laboratory, www.fueleconomy.gov; Figure 2.18: Courtesy of General Motors UK; Figures 2.19 and 2.20: © James Warren.

Chapter 3

Text

Box 3.4: Extracts from 'Case Study: Plymouth Hospitals NHS Trust', *Making Travel Plans Work: Case Study Summaries*, July 2002, Department of Transport. Crown copyright material is reproduced under Class Licence Number C01W0000065 with the permission of the Controller of HMSO and the Queen's Printer for Scotland;

Box 3.5: Extracts from 'Case Study: Nottingham City Hospital NHS Trust', *Making Travel Plans Work: Case Study Summaries*, July 2002, Department of Transport. Crown copyright material is reproduced under Class Licence Number C01W0000065 with the permission of the Controller of HMSO and the Queen's Printer for Scotland.

Figures

Figures 3.1, 3.3, 3.4, 3.5, 3.6 and 3.8b: © Marcus Enoch; Figure 3.8a: © Sally Cairns; Figures 3.2, 3.7, 3.8c and 3.9: © Stephen Potter.

Chapter 4

Text

Boxes 4.8 and 4.9: Energy Efficiency Best Practice Programme (2002) *A Travel Plan Resource Pack for Employers*. Crown copyright material is reproduced under Class Licence Number C01W0000065 with the permission of the Controller of HMSO and the Queen's Printer for Scotland.

Tables

Table 4.2: Energy Efficiency Best Practice Programme (2002) *A Travel Plan Resource Pack for Employers*. Crown copyright material is reproduced under Class Licence Number C01W0000065 with the permission of the Controller of HMSO and the Queen's Printer for Scotland.

Figures

Figures 4.1, 4.3, 4.4 and 4.5: © Marcus Enoch; Figure 4.2: Cranfield University Press; Figure 4.6: © Sally Cairns.

Conclusion

Figures

Figure c.1: Adapted from Hickman and Banister 'Looking over the horizon', *Town and Country Planning*, Vol. 75, No. 5, May 2006. Town and Country Planning Association.

Figure c.6: After Laherrère, J. H. 'Forecasting future production for past discovery', OPEC Seminar, 28 September 2001.

Index

A

B

C